Globalization, Modernity, and the City

We live in a world of big cities. Urbanization, globalization, and modernization have received considerable attention but rarely are the connections and relations between them the subjects of similar attention. Cities are an integral part of the network of globalization and important sites of modernization.

Globalization, Modernity, and the City weaves together broad social themes with detailed urban analysis to explore the connections between the rise of big cities, the creation of a global network, and the making of the modern world. It explains the growth of big cities, the urban bias of global flows, and the creation of metropolitan modernities. The text develops broad theories of the subtle and complex interactions between urbanization, globalization, and modernization in a sweep of the urban experience across the globe. Thematic chapters explore the making of the modern city in profiles of the growth of urban spectaculars, the role of *flânerie*, the traffic issues of the modernist city, recurring issues of urban utopias, and the rise of the primate city.

Detailed case studies are drawn from cities in Australia, China, and the US. Urban snapshots of cities such as Atlanta, Barcelona, Istanbul, Mumbai, and Seoul provide a truly global coverage. The book links together broad social themes with deep urban analysis. This well-written, accessible, and illustrated text will appeal to the broad audience of all those interested in the urban present and the metropolitan future.

John Rennie Short is Professor of Geography and Public Policy at the University of Maryland (UMBC), US.

Routledge studies in Human Geography

This series provides a forum for innovative, vibrant, and critical debate within Human Geography. Titles will reflect the wealth of research that is taking place in this diverse and ever-expanding field. Contributions will be drawn from the main sub-disciplines and from innovative areas of work that have no particular sub-disciplinary allegiances.

Published

Forthcoming

Globalization, Modernity, and the City

John Rennie Short

Routledge
Taylor & Francis Group

LONDON AND NEW YORK

First published 2012
by Routledge
2 Park Square, Milton Park, Abingdon, Oxon OX14 4RN

Simultaneously published in the USA and Canada
by Routledge
711 Third Avenue, New York, NY 10017

Routledge is an imprint of the Taylor & Francis Group, an informa business

British Library Cataloguing in Publication Data
A catalogue record for this book is available from the British Library

Library of Congress Cataloging in Publication Data
Short, John R.
    Globalization, modernity, and the city/John Rennie Short.
    p. cm.
    1. Urbanization—Case studies.   2. Globalization—Case studies.
    I. Title.
    HT119.S3 2011
    303.48′2091732—dc22   2011004017

ISBN: 978–0–415–67692–2 (hbk)
ISBN: 978–0–203–80886–3 (ebk)

Typeset in Times New Roman by
Florence Production Ltd, Stoodleigh, Devon

Printed and bound in Great Britain by
CPI Antony Rowe, Chippenham, Wiltshire

Contents

Case studies

Figures

Tables

Preface

I grew up in a small village in a very rural part of Scotland. It was perhaps inevitable, then, that I would have an interest in the urban. Cities, especially big cities, seemed wonderfully exciting when I was growing up, full of mystery and promise, intoxicating, transgressive, with a hint of danger and a whiff of excitement. That naïve fascination has stayed with me throughout my academic career as I have explored different facets of the urban experience from an initial interest in the dynamics of housing markets to more recent concerns encompassing globalization and environmental issues. For the past fifteen years, I have been working on exploring the relations between the city and globalization. It began with a paper in 1996 that looked at the "dirty little secret" of world cities research, namely its shaky empirical foundations. This led to more detailed empirical analysis of the global urban hierarchy—the publication in 1999 of *Globalization and The City* with my colleague Yeong-Hyun Kim as well as other articles and books, including *Global Metropolitan* (2004) and *Liquid City* (2007)—and the start of a longstanding relationship with the University of Loughborough's Globalization and World Cities (GaWC) Research Unit.

In this book I want to look again at the city-globalization nexus. In particular I want to discuss the unfolding spatial consequences of globalization, the globalizing effects of urbanization, and the impacts of these trends on an understanding of modernity. I draw upon a number of articles and books written in recent years addressing specific issues but always conceived as integral parts of a broader goal to unravel the subtle and complex relations between the city and modern world. The work is part of the spatial turn in our understanding of social and cultural issues. The original title of this book, *Big City Small World,* encapsulates the ideas of an emerging socio-spatial dialectic that consists of, on the one hand, a globalizing world dominated by the urbanized world of large metropolitan regions and, on the other hand, a process of metropolitanization very much marked and shaped by a modernizing globalization. The adopted title summarizes this awkward sentence.

In Part 1, "Global generalities," I set the scene by defining what I mean when I use the terms *third urban revolution, late modern wave of globalization* and *metropolitan modernities*. In Part 2, "Thematic specificities," chapters focus on broad interurban trends and issues. I focus on a number of subthemes that intersect with the nexus of globalization, urbanization and modernization: the city as a stage for modern spectaculars that promote and embody globalization (Chapter 4); the modern city as lived experience in terms of *flânerie* (Chapters 5) and machine/human interactions (Chapter 6); modernity's urban utopias (Chapter 7); and the global context and economic consequences of urban primacy (Chapter 8). In Part 3, "City particularities," are the theorized case studies that examine individual cities. Here I look at Alice Springs/ Mparntwe, Shanghai and the urbanized seaboard of northeastern US (Megalopolis). In these I use the individual city as both a case study in its own right but also as a microcosm of city–global–modernity themes; the postcolonial and creative city (Chapter 9), the rapidly growing reglobalizing city (Chapter 10) and what I term the liquid city (Chapter 11). I have restricted the case studies to cities with which I have some personal experience. I also discuss specific cities in some of the case studies that accompany each chapter of the book. In some cases, my experience consists of many visits over a number of years, while in others it comprises short visits. A common interpretation is to assume the long visited as more solidly grounded than the briefly experienced. While long and repeated visits can give us a deeper knowledge, sometimes they allow familiarity to blind us to the strikingly obvious. And while the solitary, brief connections can limit us to the surface obvious, they can perhaps allow us a more direct route to the heart of the matter. I hope to have combined the benefits of both while avoiding the disadvantages of either.

The three parts represent different takes on the city–globalization–modernity nexus, cross-sections through broad ideas, specific themes, and particular cities. Part 4, "Urban Utopias," contains the final chapter, more of a postscript, where I consider the prospects for building a more humane city.

A number of chapters grew from collaborative work with colleagues. Kathryn Kramer collaborated on the work that forms the basis for Chapter 5 and Luis Mauricio Pinet-Peralta worked with me on the papers that form the basis for Chapters 6 and 8. I am grateful to both of them for their intellectual generosity. A number of chapters grew out of invitations and work published in academic journals. Chapter 4 has two origins. The section on World's Fairs grew out of an invitation from Peter Taylor and Ben Derudder for a contribution on globalization and urban mega-events while the material on the Summer Olympics is drawn from an article published in 2008 in the journal *City*. Chapter 6 is a major revision and shortening of a paper that first appeared in 2010 in the journal *Mobilities*. Chapter 7 owes its origins to an invitation from Frank Eckart to deliver a paper at a conference held in Weimar in 2009 to celebrate the ninetieth anniversary of the founding of the Bauhaus. Chapter 8 is a revised version of a paper that appeared in 2009 in the journal *Geography Compass* and Chapter 9 is a revision of a paper published in 2011 in the *Annals of Association of American Geographers*. Chapter 10 arose from an invitation from the Mayor's department of Shanghai to give a paper at the 2007 International Symposium on Regional Development of the Yangtze River Delta. Faculty and students at the Department of Geography at East China Normal University allowed me to extend my stay and undertake fieldwork in the city. Chapter 11 arose from an invitation from Sophie Watson and Gary Bridge for a contribution to the second edition of *Companion to The City* as well as an invitation from Toni Tulla to give a seminar in 2009 to the Department of Geography at the Autonomous University at Bellaterra, Barcelona. Chapter 12 originates in an invite to the International Conference for Urban Humanities in Incheon South Korea in 2009. The subject of the conference was how we can build a more human city.

I also visited many cities, often aided by great guides and good friends: Rosa and Toni Tulla in Barcelona, Ki Suk Lee in Seoul, Zhuang Peijun Paige in Shanghai, Rob Freestone in Sydney, Grzegorz Weclowowicz in Warsaw, and Frank Motz of ACC-Galerie in Weimar, to name just a few. The material presented here is based on the hard work and gentle kindness of many friends and colleagues.

Part 1

Global generalities

1   The Third Urban Revolution

In this book I want to look at the connections between three dominant social trends: the growth of cities, the creation of a global society, and the construction of modernity. Urbanization, globalization, and modernization have received considerable attention but rarely are the connections and relations between them the subjects of similar attention. I want to focus on their interconnectedness, the subtle as well as the surprising links between the three. I will develop ideas concerning what I have termed the Third Urban Revolution, the Late Modern Wave of Globalization and metropolitan modernities. Let us begin in this chapter with some definitions and provisional descriptions of the Third Urban Revolution.

May 23, 2007 marks a demographic milestone. On that date the world's population became more urban than rural. For the first time in human history more people lived in cities than in the countryside. The shift marks a culmination of two centuries of continued and increasing urban growth. As late as 1800 only three out of every 100 people lived in cities. A little over 200 years later the figure was more than fifty. Our world is increasingly a world of cities and the urban context is now the dominant human condition.

Urbanization is often described as the geographical redistribution of population from rural areas to urban areas. However, it is not just the spatial reorganization of society but also the social reorganization of space. Cities incubate social and political change. They are, as Fernand Braudel (1981, p. 479) notes, "like electric transformers. They increase tension, accelerate the rhythm of exchange and constantly recharge human life."

Even when cities were small and demographically insignificant, they played a disproportionate part in social development. The first cities that appeared from the fifth millennium BCE witnessed the emergence of writing, political centralization, and organized knowledge. To take just one example: organized religions, with their gods and priests and rituals to make sense of the mysteries of life, grew out of the need to justify and legitimize the social hierarchy in the early cities. When urban empires expanded and came into contact with other urban empires, there was both creative and destructive cultural collision. Karen Armstrong (2006) updates the idea of a pivotal "axial age" from 900 to 200 BCE when the great religious traditions of the world came into being across the globe—Confucianism and Daoism in China, Hinduism and Buddhism in India, monotheism in Israel, and philosophical rationalism in Greece. Their universal messages of empathy and compassion grew out of the collective experience of living in cities as well as their awareness of people in other cities and regions. Do unto others as you would have done unto yourself is the only livable code for city dwellers living cheek by jowl.

The early merchant city—examples include Amsterdam, Bruges, Florence, and Venice—gave us the development of international trade and commerce but also the creation of a commercial society where private interest was regulated, collective rules were established, and civic communities were forged. The people of the merchant cities helped create the notion of the public realm and civic society. The largest and most prestigious building in Amsterdam during the city's golden age of mercantile activity, let us not forget, was the Town Hall (see Figure 1.1).

The industrial city was the crucible of early industrial capitalism, giving us factory life and class struggles. It was the setting for a working class as much for itself as in itself. Marx's great three volume book, *Capital*, is a work of theory based firmly on the experience of the industrial cities of nineteenth-century Britain.

Again and again, the city, in its various emblematic forms, creates the opportunities, contexts, and laboratories for new social developments, innovative political developments, and far-reaching economic changes. Cities are the accelerants of social, economic, and political change.

There are two dominant narratives that deal with the weight of this urban consequence. The first is the city as emancipatory vehicle. When Sir Peter Hall (1998, p. 7) writes that the most cosmopolitan cities, despite all their problems, "have throughout history been the places that ignited the sacred flame of the human intelligence and the human imagination," he repeats an old idea that cities help to liberate us from the past. In this reading, the city is the setting for the full realization of human potential and the launching pad for freedom. This is the city as the leading edge of a progressive history. Another, alternative reading sees the city as a site of compulsion. The authoritarian city is the counterpoise to the liberation city. Cities have an authority embedded in them that ranges from the more obvious forms of control and surveillance to the internalized, the guards, visible and internalized, that limit our movement and shackle our imagination. The city is of course too complex to be easily summarized by either characterization. The liberation city and authoritarian city are best visualized as polar ends of a shifting dynamic continuum. Occasionally the liberation city spurts into life: riots in Paris, marches on Washington, people power in Manila. This is where and when the city is a theater for strikes and sit-downs and demonstrations. At the other end are the policeman's cosh, the fire hose, the prison, and the more subtle but effective dominant ideologies that marginalize and smash alternatives. And in between is the vast banal reality that cities are places of quiet normality, rarely the focus for insurrection, rebellion, or disobedience.

Figure 1.1 Amsterdam Town Hall: a monument to the urban polity.
Photo: John Rennie Short

Contemporary urbanization

Today we are in the midst of the Third Urban Revolution. The first began over 6,000 years ago and saw the first cities in Mesopotamia. These cities were less the result of an agricultural surplus and more the reflections of concentrated social power that organized sophisticated irrigation schemes and vast building projects. The First Urban Revolution, independently experienced in Africa, Asia, and the Americas ushered wrenching social changes, new ways of doing things, and new ways of experiencing, seeing, and representing the world. The Second Urban Revolution began in the eighteenth century with the linkage between urbanization and industrialization that inaugurated the creation of the industrial city and unleashed unparalleled rates of urban growth. Since 1800, urban growth is one of the most significant features of global demographic change.

The Third Urban Revolution is a complex phenomenon that began in the last half of the twentieth century. It is marked by five distinct characteristics. The first is the sheer scale and pace of change. A majority of the world's population is now urban and in many countries this new urban majority appeared in little more than a generation. The proportion of the urban population in developing countries has doubled between 1950 and the present day. In 1950 the urban population of Kenya was 340,000, by 2005 it had increased to 7,384,000. In Pakistan the urban population in 1950 was 6,473,000, and by 2005 it was 55,135,000. Some of the effects of this rapid change in culture will be explored later in our discussion of modernity and modernization. For the moment we can note that across the world, the dominant demographic trend is the growth of cities and the development of urban places.

The second characteristic is the increasing size of individual cities. Urban growth is concentrated in large cities. Take the case of Shanghai. In 1950 it had a population of around six million; by 2010 its official size approached close to twenty million. In the developing and developed world, the big cities have morphed into wider metropolitan regions. In 1900 no metro region in the world had a population greater than ten million; a hundred years later there were at least twenty and probably more. Population and economic activities have spread beyond the municipal boundaries. In the past fifty years, across the world, small towns have grown into cities and big cities have sprawled into giant metropolitan regions.

Third, this revolution exhibits a marked metropolitanization. Improvements in transport have allowed dispersal of people and activities away from the tight urban cores of the pre-industrial and industrial cities. Large metropolitan regions rather than individual cities are the new building block of both national and global economies. Three giant urban regions in Asia Pacific, Bangkok (11 million population), Seoul (20 million), and Jakarta (20 million), have between 35 and 25 percent of all foreign direct investment into their respective countries and constitute between 20 percent and 40 percent of respective national gross domestic product. In China, for example, the three city regions of Beijing, Shanghai and Hong Kong constitute less than 8 percent of the national population, yet they attract 73 percent of the foreign investment and produce 73 percent of all exports. China is less a national economy than the aggregate of three large metropolitan economies. Recent work looking at city regions in the US has identified ten megapolitan regions, defined as clustered networks of metropolitan regions that have a population either of more than ten million or that will exceed that number, on current growth projections, by 2040 (Lang and Dhavale 2005). Collectively they constitute only 19.8 percent of the nations' land surface yet comprise 67.4 percent of the population. I examine one of the regions in Chapter 11. Large city regions are the main centers of global economic activity.

Fourth, this urban change is a global phenomenon. The First Urban Revolution was concentrated in the fertile plains of only a few river basins. The second was restricted to cities in countries undergoing rapid industrialization. The Third Urban Revolution, in contrast, is truly global. Metropolitan growth is evident across the globe and patterns of individual city growth and decline are linked to global redistributions of economic activities. The global shift in manufacturing from the West to new centers of industrial production has urban consequences. Some of the slowest growing cities of the past decades are places such as Liverpool and Leeds in the UK and Buffalo and Pittsburgh in the US. Their relative decline is connected to their loss of manufacturing activity and the decline of the industrial base. Meanwhile cities such as Shanghai and Seoul, as first manufacturing and then service employment, grew substantially (see Figure 1.2). The tighter global linkages ensure that global shifts in economic activity are quickly reflected in patterns of urban growth and decline. Urban landscapes became devalorized and revalorized at an often bewildering pace: in much of the developed world central cities have characteristically become sites of new urban spectacle; inner cities are pock-marked by sites of gentrified renaissance as well as stubborn poverty; inner suburbs show the first inklings of decline while exurban development continues apace as gated communities and mixed use developments sprawl into the former countryside. In the developing world massive rural to urban migration feeds the creation of peripheral settlement on the edge of cities and fast growing informal economies. Urban growth seems inexorable around the world, just as urban decline seems unavoidable throughout the globe.

Fifth, large cities are now part of a new spatial assemblage of connections between the global, the nation-state, and the city. The territorial unity of nation-states is being undermined by changes in national policies as well as global trends. The encouragement of the private market and the decline of the Keynesian state is leading to marked regional inequality and difference. In many countries around the world the difference between urban and rural, big city, and small town, expanding and declining economies is reinforced by the lack of national equalization policies. The "national" economy is something of a statistical fiction, an averaging of different urban regions. Economic globalization is aiding the creation of informal city-states within different countries. Sydney in Australia, for example, is developing as

Figure 1.2 Songdo International City, South Korea: part of the massive and rapid change in the Third Urban Revolution.

Photo: John Rennie Short

Australia's "global city." Its appearance and the life experiences of its citizens are becoming more like those in San Francisco, London and New York than those in the outback or small town Australia. The ability of Sydney to reposition itself in the global economy allows the city to become separated off from the rest of the country. Globalization is aiding in the creation of de facto city-states, the control points of a global economy. Globalization is not so much undermining the nation-state as restructuring it. The nation-state is being redifferentiated according to success and failure in dealing with and attracting global capital. Certain parts are becoming more successful, some less so; in effect, the nation-state is being broken up into degrees of connection with the global economy. In the wake of a widespread withdrawal of state intervention, the result is growing difference within the nation-states, increasing similarity between certain cities in different countries, and the rise of de facto city-states whose economic fortunes are tied to their global connections as much as to their national location.

The Third Urban Revolution embodies and reflects global trends whether it is worldwide shifts in economic activities, leading to new industrial growth in some metropolitan regions and the deindustrialization of older regions, or increased international immigration, in terms of the emergence of cosmopolitan cities and urban immigrant gateways. There are connections between this latest round of urbanization, globalization, and modernization that I will now explore.

CASE STUDY 1.1 **Istanbul**

To say that Istanbul is situated between East and West is not a trite cliché, but a simple statement of geographical reality. The city sprawls across the narrow sea passage between Europe and Asia.

Like all maritime cities—it fringes not only the Bosporus, but also the Golden Horn and the evocatively named Sea of Marmara—it has an expansive luminosity, a salt-tinged clarity and sea-breezed airiness. Even with seventeen million people and busy traffic, it still manages to shine and sparkle.

The city straddles both continents and embodies the varied and long contacts between East and West. Founded in 330 by the Emperor Constantine as Constantinople, it became the capital of a Byzantine Empire that stretched throughout the Eastern Mediterranean, covering parts of Europe as well as Asia. As an important repository of the legacy of Greek classical thought and a transmission point for the intellectual flowering of Arabic and Persian scholarship, Constantinople stayed lit when Western Europe sank into the dark ages. The Byzantines kept the sparks of culture alive that subsequently caught fire in the Renaissance of Western Europe. There were also more destructive exchanges. The city was pillaged in the Fourth Crusade, its treasures looted and taken back to Western Europe. The churches of Venice are filled with the theft.

When it became the capital of the Ottomans, it was the center of an empire that embraced the continents of Africa, Asia, and Europe. At its furthest extent, the city's power reached to the very gates of Venice. Under the longest reigning Sultan, Suleiman the Magnificent (r. 1520–1566), not only was the empire extended, but also the city was adorned with new mosques and an early form of multicultural tolerance was practiced. The city is still filled with the most beautiful of religious architecture including the intimate Church of St Savior in Chora, the serene Hagia Sophia, and the stunning Blue Mosque. The city skyline as it stretches across the hills is punctuated with minarets and towers, Ottoman mosques, and Byzantine churches.

As a center of two transnational empires the city was a global city. Charles Parker (2010) writes of the global integration of space as the defining feature of early modernity. From 1400 to 1800 empire building across the globe—including the Chinese, Ottoman, Mughal, and Safavid as well as the Spanish and British—created international markets and global exchange networks, the movement of people, the spread of new technologies, the diffusion of cultures, and the transmission of religion and scientific practices. The result was a tighter integration of global space, a space pulled especially tight in cities such as Constantinople/Istanbul.

The names *Byzantine* and *Ottoman* cover processes rather than fixed objects. The Byzantine Christians grafted their religion onto pagan roots, and their later art reflected contact with the East as well as encounters with the West. The icons produced in the later Byzantine Empire have a more naturalistic representation, reflecting contact with the art of the Western Renaissance. Constantinople/Istanbul was a city where Arabic and Turkish, as well as Greek and Italian, could

Figure 1.3 Grand Bazaar in Istanbul: completed in 1461, it is one of the largest and
oldest covered markets in the world; a forerunner of the Paris Arcades
and today's indoor malls.

Photo: John Rennie Short

be heard in the streets. As the Ottoman Empire waned, there was a growing admiration for an ascending West. In 1856, the Sultan abandoned the Topkapi Palace for a new neoclassical palace at Dolmabahçe, which could look just as at home on the banks of the Seine or the Thames as it does on the Bosporus. The encounters continue. Today the Grand Bazaar, first founded when the Ottomans gained control of the city in 1453, is filled more by foreign tourists than native Turks (see Figure 1.3).

Istanbul is less a fixture and more an ongoing process, a crucible where East and West are reshaped. Turkey is a longstanding, vital member of NATO and continues to forge further links with the West. Its application to join the European Union is just one of many ventures reaching to the West. There is also a more recent religiosity that derives from the Islamic world. To walk the streets of Istanbul is to see women dressed in the most contemporary fashions as well as in the hijab. Islamic fundamentalism in Turkey is more an invented tradition of relatively recent vintage than a "pure" historical artifact; in part, a result of the massive rural–urban migration from the villages of Anatolia to the peripheral settlements where religion provides a softer landing to the modern world and religious organizations fill the gap in the welfare functions of the state. The fundamentalist and the secular represent not the past and the future, respectively, but alternative conceptions of the present: Istanbul's geographical reality stages these competing modernities in the form of tension between the comfort of an invented past and the attraction of an imagined future. Founded as a secular republic, modern Turkey is still situated between East and West, and at the heart of this unfolding drama is the city of Istanbul.

CASE STUDY 1.2 **Flash and intimate urbanism**

There are two types of urbanism. There is *flash urbanism*; this is city as urban spectacular, building as starchitecture, the big statement, the grand flourish. For the past hundred years the persistent theme of flash urbanism is the construction of the tall building. Cities predicated upon flash urbanism include the oil rich states of Abu Dhabi and Dubai. Here, a combination of lots of money and a desperate need for world respect and global recognition leads to an architectural frenzy. Abu Dhabi has a Zaha Hadid-designed performing-arts center and a Frank Gehry-designed Guggenheim museum. Its central business district is a concrete forest of new tall buildings. Currently, Dubai boasts the world's tallest building, the 160-floor Burj Khalifa and the 77-floor Emirates Park Tower, the tallest hotel in the world. Tall buildings are spectacular and as urban skyline they create a visual signature of a globalizing modernity. "Everything in Dubai," notes the architectural critic Paul Goldberger (2010, p. 64), "is a kind of visual spectacle intended to make you gawk . . . to make sure the world knows who you are."

But what is life like at the bottom of these buildings? In the Arab Emirates, streets are rarely filled with people. Almost a quarter of a million laborers work to construct Dubai's urban fantasy of global city rather than a city of enriched human experience. In Pudong the buildings that stand so tall rarely link with each other or the surrounding urban fabric. Too close to drive, the locals complain, and too far to walk from each other. Flash urbanism is about surface appearance, about impressing, about visual rhetoric. It is about the now.

The now can soon become the unrealized future. Flash urbanism is often fueled by long credit lines that are quickly cut and reduced during downturns. Cities around the world are filled with abandoned projects and half-finished buildings such as Gran Torre Costanera, which was planned to be the tallest skyscraper in South America. Of the planned sixty storeys, only twenty-two were completed before the project folded. The Russia Tower in Moscow was planned as the tallest building in the world on its completion in 2012. The project was canceled in 2009 and the site lies empty and forlorn. The city of Bangkok is now a forest of abandoned towers when speculative construction abruptly stopped in the 1997–98 Asian financial crises. Almost 320 high rise towers, some of them arching up to forty-seven storeys, lie abandoned and rotting in the tropical humidity, giving mute testimony to the vagaries of the business cycle and the failed ambitions of speculative developments. Flash urbanism can fizzle and flame out as well as arise and dazzle.

There is another, quieter form of urbanism that I will refer to as a more *intimate urbanism*, smaller scale, more linked to the human body and its locomotion. An exemplar is the urban fabric of Dutch cities in their walkability, their deference to the bicyclists, the reliance on public transport, their quieter, more intimate feel. The Dutch prize what translates as coziness, and this is obvious in their built form. The Netherlands is not known for tall impressive buildings as in Abu Dhabi, Dubai, or Shanghai. It is better known for the intimacy of its streets, the coziness of its public space, the human scale of its cities.

Flash urbanism and intimate urbanism. They are not so much separate but two poles of a continuum. Sometimes the flash can turn with age and use and custom into the intimate as a city incorporates the brash and new into the old and familiar. The Town Hall in Amsterdam, a monument to civic pride and civic society, was at one time one of the largest municipal buildings in Europe. Now it is a much-loved landmark in Dam Square. And even in cities such as Dubai foreign laborers create more intimate places of encounter and interaction at the edges of large empty public spaces, a subtle and embodied form of resistance to their marginalization by the flash urbanism that they themselves build (Elsheshtawy 2008). Flash urbanism reflects the "make-an-impression-now": it is the architectural equivalent of instant gratification. It took less than twenty years to turn rice paddies into Pudong's skyline. Intimate urbanism takes a while to develop. Dutch urbanism emerged over centuries. While it is easy and quick to follow fashion, it takes longer to find out what works as a livable city. I love to visit flash cities but I prefer to live in more intimate cities.

———————————

2   The Late Modern Wave
of Globalization

Globalization is neither recent nor new. The year 1492 marks the date when the separate hemispheres were linked permanently to form a unitary globe. This was the first wave of globalization. The movement of ideas, people, capital, goods, and viruses all were part of the Columbian exchange that stitched the separate hemispheres into a single unit. Colonial powers deepened the connections, bringing new language and customs while taking away precious goods and wealth. This world was an unequal one with global participation always limited, partial, and unbalanced. In the late nineteenth century another wave of globalization widened and deepened the global connections. New regulatory modes were added as major powers sought to expand their reach and negotiate with their rivals. The second wave of globalization, the imperial/colonial age from 1880 to 1918 saw the partition of Africa, the creation of international organizations and institutions, the final wresting of the American West from indigenous peoples, and the creation of colonial economies and industrialized nations.

The second wave came to an end in the interwar period of the twentieth century when the Great Depression quickly supplanted economic growth, global trade was reduced, foreign immigration restricted, and protectionist policies were adopted. The resulting economic chaos paved the way for a new era of global economic regulation. During the closing years of World War II, a new regime of regulation was established to stabilize the international economy. The experience of the Great Depression, when stagnation was reinforced by national governments' pursuit of protectionist policies, highlighted both the dangers of unregulated international trade and the mutual benefits of managed international cooperation. The Bretton Woods agreement was signed in 1944; its general aim was to maintain national economic sovereignty for the capitalist developed world with new rules to foster cooperation between countries. The World Bank, International Monetary Fund (IMF), and a system of fixed exchange rates were all established. The fixed exchange rates, which effectively kept currencies pegged at fixed values with the US dollar acting as convertible medium of currency, lasted until 1971 when a more liberal and deregulated global financial system emerged, ushering in the present wave of globalization. This is the third wave of globalization that I will term the Late

Modern Wave of Globalization to highlight its temporal and more complex relationship to modernization than the preceding waves. To be more accurate, it is a wave of reglobalization as it courses through the structures created by the previous two waves. These reglobalizations leave an urban mark, sometimes even in the same city. In Chapter 10 I consider Shanghai as a city shaped by the previous and present waves of globalization, its form and character the result of cumulative reglobalizations.

This Late Modern Wave has three characteristics that I will subsume under the general themes of economic, political, and cultural globalization and examine through the contested propositions of a flat world, a neoliberal world, and a homogeneous world. I draw heavily on previous work (Short 2001; 2004a).

Economic globalization: a flat world?

The world is now flatter than it has ever been. By flatter I mean that a global economy has emerged with easier and hence quicker economic links across the globe. The icon of the flat world is the shipping container. First developed in 1956 and used effectively in the Vietnam War to ship goods from the US to Southeast Asia, the container allows the transportation of trailer-sized loads from ships to trains and trucks without breaking bulk easily and cheaply around the world (Levinson 2006) (see Figure 2.1). It is not just that the shipping container created economic globalization, but that globalization created the container. The container both flattened the world and emerged from this flattening. A series of decisions from 1968 to 1970 by the International Organization of Standardization created global standards of size, markings, and corner fittings. The result was a truly global form of cheap transportation. Today, close to twenty million containers make over 200 million journeys per year. Containers loaded with goods in a Chinese factory can be driven by truck to a rail depot where they are transferred to a port where they are loaded onto ships that sail across the Pacific and then unloaded easily and quickly onto trucks, say in Long Beach, California, and transferred to regional distribution centers, from there to be sent to local stores for purchase by consumers. The ease and low cost of moving this international cargo enables mass manufacturing to relocate toward the cheaper labor cost areas of their world.

There has long been global trade, but for centuries the high cost of transport restricted it to the luxurious—goods or things simply unavailable in the local or national economy. Now the steep fall in transport costs makes available everyday things from around the world. The plummeting transport costs infuse the global trade with even the small, the mundane, and the cheap. Blueberries grown in Chile, T-shirts manufactured in Vietnam, cheap plastic toys made in China, or white wines bottled in New Zealand can all be purchased in the same store in Syracuse, New York. The global trade in goods is an important part of our everyday lives. And global trade now has the power to influence, affect and even destroy local and national economies.

Figure 2.1 Container boxes on a ship in the Panama Canal. The Canal, completed in 1914, with new locks planned for 2015, is part of global space–time convergence.

Photo: John Rennie Short

Distance and transport costs are no longer the barrier they once were to global competition in this Late Modern Wave. A flatter world is a smaller world.

The flattening was not simply a result of low transport costs that enable relatively cheap products to be made in one country and sold in another halfway around the world: it was further lubricated by the de-skilling of manufacturing. As jobs were routinized and de-skilled they were capable of being offshored to regions with little manufacturing history. Production could be prized away from the pools of skilled and higher paid labor in traditional manufacturing regions. This shift had two consequences. The first was the rise of a new working class in the newly created industrial regions. A female working class emerged as young women filled many of the new jobs. It is still too early to tell whether this class in itself will become a class for itself. But already new gender identities are forged in such places as the maquilodires along the US/Mexican border or in the industrial cities of coastal China. The era of cheap labor in China, for example, may be coming to an end as the population ages, the supply/demand relationship favors labor more than capital and especially as labor organizes a potent force in the large coastal cities.

The second consequence is the decline in the economic and political strength of the organized working class in the old industrial regions of Europe and North America. From the end of World War II to the mid 1970s, organized labor in the developed world was able to increase its real wages and in some countries promote a broader social agenda that shifted WARFARE-welfare states to WELFARE-warfare states. From the mid 1970s the decline of the manufacturing base, the loss of relatively well paid industrial jobs and the general weakening of organized labor resulted in a general decline of the bargaining power of labor. As economic and political power shifted from labor to capital, the cascading effects were a decline in real wages and the replacement of a social welfarism and economic Keynesianism with neoliberalism, privatization, and deregulation.

These moves were resisted and contested and the balance of forces varied by country and so we have differential patterns of greater or lesser degrees of residual welfarism with resurgent neoliberalism. At one end are the Nordic social democracies where a popular support of welfare policies has muted neoliberalism. However, the continuing commitment to high welfare standards has its own problems. With an ageing population and shrinking working population, a fiscal crisis emerges as taxation must be deep and wide in order to sustain the programs, which increase in costs every year. These democracies face the problem of maintaining legitimacy in part through the commitment to welfare on the one hand with paying for the programs on the other hand. High taxation may undermine international competitiveness as well as undercut the work ethic. If benefits are high and so are taxes on wages, then the need to work is lessened. In some of these democracies, many of the costs are borne by the younger generation and the immigrants. With more secure employment contracts for more established workers, the younger workers suffer higher unemployment. And immigrants can pay taxes but are often denied the social benefits. The sharp elbows of the middle-class crowd out low-income households for scarce or rationed welfare resources. Such inequities of welfare democracies can refine and redefine the notion of welfare along the lines of "I obtain the due benefits of citizenship, you receive welfare payments." Issues of citizenship and who is responsible for paying taxes and receiving benefits also become crucial questions for domestic policies; they also provoke deeper discourses of national identity. The fiscal fissures of a welfare state reveal themselves most strikingly during economic downturns and when global competitiveness becomes more marked. They bring to the surface the issues of who benefits from welfare and who pays for it, issues that are exaggerated if there is perception of high foreign immigration levels and an increase in the amount of the foreign born population. The fissures reveal the deeper cleavages in society along the lines of class, gender, generational cohort, and citizenship status.

At the other end of the spectrum is the US where the commitment to welfare is more muted and a more market orientated police culture dominates. Here, organized labor traditionally fought more for industry-based welfare based on occupation rather than a society-wide welfare based on citizenship. This model worked well in the immediate postwar period up to the early 1970s as manufacturing employment increased. US manufacturing and industrial supremacy created a vast middle class. But with the demise of the industrial base, the shift in power from labor to capital was not softened by generous welfare. The US saw a great increase in income inequality from the 1970s as the middle class were squeezed while real wages for working people stayed flat or declined; it now has the highest income inequities in the developed world. In 1975 the Chief Executive Officer (CEO) in a private company made the equivalent of thirty-six families earning the US median income. Twenty years later they made the equivalent of 133 families earning the median income. In 1990 the average CEO to average worker pay ratio was 42, by 2006 it was 364, falling only to 319 two years later in the middle of the Great Recession. The wage gap between skilled and unskilled workers has also increased. During the long boom many workers could obtain relatively high wage jobs in the industrial sector without the need of a college education. With a more globalized economy, these jobs moved offshore, or technological improvements in productivity meant there were fewer of them. The result is less demand for unskilled labor and a consequent decline in income. In 2006 the median average income of white males was $40,432. For those with Master's degrees, it was $77,818. Racial and gender inequalities overlie the skill gap. For Hispanic women the respective figures were $20,133 and $47,301.

The big box retailers also reinforced the world's flattening. Traditionally, manufacturers made things, then sold to retailers who sold them to customers. Retailers were simply in the middle, buying goods from producers, then selling goods to customers. But with the rise of the big box retailer and a more flexible production system, with many small companies producing small runs of goods relatively quickly, power shifted from the large number of small manufactures to the small number of larger retailers. The very large retailers have such enormous power that they can force producers to compete among each other to meet strict cost limits. Manufacturers are willing to do this because access to a big retailer assures a large customer base. The lower price per item is offset by the higher turnover of items.

Take the case of Walmart, which in 2010 had almost 8,500 stores in fifteen countries. Walmart uses its enormous leverage to drive down costs of manufacturers who supply these stores. Walmart reduces the price of goods by relying on producers from low wage countries thus pushing North American producers to reduce their costs. In this flatter world a number of North American producers have offshored their production to reduce costs in order to sell to Walmart. Walmart and stores like it are a major cause for the shift in manufacturing to lower-wage regions. One clothing retailer, Hennes and Mauritz (H&M), a Swedish company with stores in Europe and North America, keeps its prices low by contracting in low-wage areas of the world. Almost 900 factories are used to produce a constantly changing design portfolio. The company has been successful in keeping inventory low; the Just in Time production system ensures that goods are made to meet demand. Stores often receive daily supplies. High turnover means that profits can be made through selling many items rather than one; hence the price of individual items can be reduced, which in turn aids turnover.

Walmart uses its market power to reduce costs, leading to more offshoring and more imported goods from cheap labor regions, which in turn leads to fewer jobs and lower incomes in the formerly richer regions and consequently lower wages, forcing more people to seek deals at Walmart, which means . . . you get the general causal picture. While big box retailers reduce costs and increase productivity, there is also a broader narrative of the downward spiral of job loss and wage reduction.

Production chains that stretch and snake their way across the globe looking for competitive advantage now link consumers in the developed world with producers in the developing world. Goods move along the seemingly friction-less distance of these chains. The process is driven by competition. At one end, consumers seek greater value and retailers, competing with one another to provide the best value, force producers into cheaper labor areas and more efficient production zones. At the other end of the chain, producers need to keep finding ways to reduce costs, improving techniques, using cheaper labor. As distance shrinks, relocation to ever-cheaper labor areas continues to flatten the world.

A flatter world allows regional and national companies to become trans-national in their operations. Peter Dicken (2003) identifies four ideal types of transnational corporations (TNCs): *multinational*, with decentralized organi-zational structure, that operates as a series of semi-independent entities; *inter-national*, controlled from one central headquarters with overseas operations as appendages; *global*, a centralized hub that implements parent company strategies to reach a global market; and *integrated network*, a network of complex, globally integrated operations. We can get some idea of the range by considering three TNCs: Nike, Starbucks, and Toyota.

Nike is a sportswear and equipment company that exhibits characteristics of the integrated network TNC. Here I draw heavily upon some earlier works (Short 2001, 2004a). In 2009 its revenue was $19.1 billion. The founder is Phil Knight who was a member of the University of Oregon track team in the late 1950s. Knight went to Stanford Business School and, when faced with a term paper, he developed the idea that low-cost Japanese running shoes could find a market niche in the US athletic shoe market. Knight did not pursue the idea immediately—he became an accountant in Oregon—but on a trip to Japan in 1963 he picked up a pair of Tiger running shoes. He showed them to his old track coach Bill Bowerman who thought they were better than Adidas shoes. They invested one thousand dollars in a thousand pairs of Tiger shoes and sold them at high school track meets. It was the beginning of a lucrative connection. Bowerman would send new designs to Japan and new shoes would be made, shipped back to Oregon and sold at track events. By 1969 the annual sales were almost a million dollars. Knight and Bowerman were selling the shoes under the Japanese brand names. In 1971 Knight decided it was time for a separate identity. The shoes were named after the Greek goddess of victory, Nike. The swoosh, the fat check mark icon, was designed by a Portland design student in 1972. Annual sales that year were $3.2 million. By 1980 they were $270 million, and one out of every three Americans owned a pair of Nikes. When the stock went public in 1980, Phil Knight's estimated worth was $100 million; by 1993 it was over a billion dollars. By 1998 Nike was making ninety million shoes per year and generating annual revenue of $9.6 billion. Ten years later it was generating twice that revenue as it branched out to golf, soccer, and many different sports with a range of sporting apparel. Nike relies heavily on advertising, often signing big name sport stars to multimillion contracts.

Selling shoes was Nike's initial business and still an important part of its operations. Making shoes is dirty, dangerous, and difficult. Initially, Nike shoes were made in Japan. After World War II, Japan had started out on a trajectory of rapid growth in manufacturing. The Japanese perfected design improvements and the products they sold on the world market got better and better. So did their wages and working conditions. Labor costs rose. Nike signed contracts with Korean shoemakers. By the early 1980s most Nike shoes were made in Korea, and the city of Pusan became the capital of Asian shoe manufacturing. Factories sprouted up, more workers were employed. South Korea was one of the fastest growing manufacturing nations in the world. Meanwhile, during the 1970s and 1980s, 65,000 jobs in shoe manufacturing in the US were lost. By the mid 1990s a pair of Nike shoes that sold for $30 actually cost around $4.50 to make. Line workers in Korea were receiving $800 a month. But in the competitive shoe business, further cost reductions were needed. Nike could reduce their costs by getting their shoes made in China, Indonesia, and Vietnam, where labor costs per worker were only $100 a month. In Vietnam in 1998, workers at a Nike shoe manufacturing plant earned as little as $1.60 a day. In Indonesia, workers were sometimes receiving as little as 50 cents a day. Indonesia is now one of the largest suppliers of Nike shoes: seventeen factories employ 90,000 workers producing around seven million pairs of shoes. In South China the centre of shoe manufacturing is the city of Guangzhou. Just outside the city, one shoe factory used by Nike makes 35,000 shoes a day.

The standard model of industrial production is often called Fordist, after the assembly-line techniques devised by Henry Ford. Large factories in fixed locations produced large batches of a limited range of goods. You could have any color Model T Ford, went the old joke, as long as it was black. Ford workers made Ford cars in Ford factories. The plants were huge, fixed capital investments. Bargaining between capital and labor thus took place against a fixed location. There is now a more flexible form of production. Nike has no factories and no Nike workers. Nike goods are made under contract by a range of manufacturers. Factories compete to obtain Nike orders and are then licensed by Nike if they are capable of making goods to strict cost and design specifications. Many of the "Nike" shoe factories in Indonesia and China are owned by Korean and Taiwanese business interests. Shoes and other sportswear are made cheaply in low-wage areas and then sold at substantial markup in more affluent market areas. The markup pays for the expensive advertising campaigns and celebrity endorsement deals.

Competition drives the system. If you are Nike, you have to compete with Adidas and others in a fast changing market where consumers are always looking for the best deals and latest designs. If you are a sportswear manufacturer, you have to compete with other factories to win the orders from companies such as Nike. If you have a shoe store, for example, you need to provide a wide range of constantly updated shoes at good prices. The intense global competition ultimately leads to better deals for consumers. The system produces the kingdom of consumer sovereignty where consumers have real choices. Good quality items are available at competitive prices in much of the rich world. In real terms, the prices of clothing and shoes have decreased. Globalization has worked in the interests of the rich consumers.

What about the workers in the new areas of global manufacturing production? Initially wages are low; that is why the companies were there in the first place, but in the first wave of newly industrializing countries, such as South Korea, wages did rise. A successful industrial economy grew and a substantial middle class was created. In the current second wave of newly industrializing countries, such as Indonesia, China, and Vietnam, the experience has varied. There has been evidence of exploitation. The publicity of low wages, forced overtime, and dreadful working conditions created a movement against sweatshop conditions in developing countries. As a high profile company Nike is singled out for some heavy criticism. Bad publicity made Nike update its Code of Conduct in 1997 to include right of free association, paying at least the minimum wage and restricting working hours to sixty hours per week. Global civil society had exercised some leverage on Nike, a company whose success is determined by its public image. A Nike company report in 1998 claimed that they ceased business with eight factories in four countries that did not meet their Code of Conduct. The negative publicity of awful conditions in factories producing apparel for Nike did lead to some changes. However, as recently as 2008, an Australian television documentary showed appalling working conditions in a "Nike" factory in Malaysia.

Toyota is an automotive company headquartered in Japan but with factories around the world and a global workforce of over 300,000. It strongly exhibits characteristics of the international and integrated network TNC. Toyota emerged in the 1930s and produced its first car in 1935. It made cars in Japan but wanted to sell them abroad. The problem was that many countries, especially those with a substantial auto industry, put up tariff walls to protect domestic producers. In order to jump these tariff walls, Toyota established manufacturing plants in selected countries. It became a multinational company headquartered in Japan but with manufacturing plants around the world. But, unlike Nike, it retained direct control of the factories producing its products. Toyota pursued a relentless drive to improve the quality and reliability of its vehicles. While US carmakers remained smug and top-heavy, Toyota, like many Japanese automakers, pursued lean manufacturing—creating more value with less work—and just-in-time production to reduce inventory and carrying costs. Over the years they established a reputation for reliable well-made cars that ranged from the budget to the luxury. A successful global brand was established, and in 2008 it became the world's largest automaker and the fifth largest company on the world with significant market shares in North America, Asia, and Africa and smaller shares in Europe. It makes around eight million vehicles a year, with manufacturing and assembly plants in twenty-seven different countries. Toyota has five major plants in the US.

Although a global company, Toyota is still anchored in Japan. The company is headquartered there, and the very senior management tends to be Japanese. As late as 2008, almost 5.1 million of the 8.5 million vehicles were produced in Japan.

The explicit desire to become the world's number one automaker was an ambitious one. This drive to global dominance is, according to many analysts, the root of its recall problems in 2009–10 when over 8.3 million vehicles were recalled for either accelerator or braking problems or both. The rapid increase in production perhaps outstripped quality control and the ability of engineers and managers to carefully scrutinize new model developments for design problems and associated safety issues. The problems were widely reported. A global company also gets global publicity when things go wrong.

The first Starbucks opened in 1971 with a small shop front store in Seattle. Its globalization is relatively recent. The first store outside the US opened in Tokyo in 1996. Starbucks now has 16,635 stores in forty-nine countries and a global workforce of 130,000. The biggest growth was in the 1990s when the company was opening a store every day. Since the recession of 2008 store closures have occurred as well as store openings. In Australia in 2008, sixty-one of the company's eighty-four stores were closed.

Starbucks cafés have remarkable similarity around the world. There are few nods to local particularity or national differences. The emphasis is on the generic. Sometimes this leads to success. Starbucks in Japan operated the same no-smoking policy as their US stores. This was unusual in Japan but that was part of its success in Japan. In some cases the generic nature also creates backlash. There was early difficulty breaking into European markets, and Italy is still a hold out—a country with thousands of coffee bars, but no Starbucks.

What Starbucks has done is less to meet a demand than to create a demand. Starbucks now embodies a global coffee culture, sometimes building on existing cultural traditions, in other places creating them anew. It has become a third place between home and workplace where socializing and tele-commuting, as much as coffee drinking occur, a meeting place as much as a selling place. While in some places it signifies a US cultural dominance, in others it is perceived as a symbol of modernity.

Nike, Toyota, and Starbucks: all global companies with recognizable global resonance. They have benefited from but also helped to create a flatter world where people in different countries wear, drive and drink the same things.

But is the world really flat? Is the smoothing of the differences enough to justify using the term "flat world"? Two counterpoints can be made. The first is the simple fact that only some of the world is flat, and by flat I mean a smoothing of the tyranny of economic distance and a lessening of the friction of national economic differences. Production chains only snake their way into selected counties. Nike athletic shoes are made from components in some countries including China, Indonesia, and Vietnam, but not in Burma, Pakistan, Somalia, or Zimbabwe and many other developing and poor countries. Starbucks has few stores in sub-Saharan Africa. There are limits to economic globalization. The cheapness of labor is not the only issue for a mobile capital. Decent reliable infrastructure, cheap transport links, the rule of law, and a stable political system are all prerequisites for long-term global shift, and much of the developing world lacks one or many of these elements. The world is not flat. Parts of it are, where distance is almost frictionless and economic transactions are easily connected to other parts of the world. But in other parts, distance is only overcome at great cost, and there are high transactional costs. We live in a misshapen world, only parts of which are flat, others curved and yet others pockmarked by disconnected ravines and inaccessible peaks.

The world's topography is more varied than flat. At the international level, there are still many countries only loosely connected to global flows of investment or global shifts in manufacturing. And even within the more connected, flatter countries, there is often a marked demarcation between the more inaccessible rural areas and the metropolitan regions where most of the global economic activity is concentrated. Much of the rural areas of the developing world are far from markets and lack the necessary vital infra-structure to ensure easy and cheap transportations—the hallmark of a flat world. The flat areas, where much of the world's global economic activities take place, are like landing strips in the middle of rugged terrain, often constituting small parts of larger countries and, in total, only a tiny part of the earth's economic surface.

The second point is that economic globalization is also about creating hierarchies as well as flat surfaces: it generates spatial difference as well as homogeneity. Economic globalization is not just the product of a consumerist culture and a competitive capitalism continually seeking to squeeze costs. Since the 1970s and the formal ending of the Bretton Woods system, there has also been a globalization of finance. Capital now flows in larger and more viscous currents across the world. In this case, the flattening of the world is also tied to the networking of the world. Flat surfaces are ideal to move goods cheaply and quickly, but in order to do business you need nodes of concentrated activity. A global system of cities, connected by flows of capital, ideas, information, and skilled personnel, "landscapes" the world into concentrated nodes of heightened connectivity and accumulated control.

In summary, while there has been a flattening of the world, a flat world is not yet a reality. Substantial differences remain. The unthinking use of the term "flat world" tells us as much about the observer as the observed. We can deconstruct the notion of a flat world by looking at the work of the main popularizer of the term, the journalist Thomas Friedman, who used the term in the title of a book, *The World is Flat*, first published in 2005. Friedman is a good writer who simplifies complex issues in breezy and easy to read prose. He presents a picture of a world flattened by technology, trade, and financial integration. He reports on the upper circuits of the globalizing world. His prose is full of such telling phrases as "as I said to the CEO of Microsoft," "on my flight to Mumbai," and "as I said to the audience at Davos." It is the experience of someone on a very comfortable perch atop an affluent lifestyle, traveling at the front of the plane, meeting with world leaders, and serviced by people from around the world. While his world is flat, he just fails to see that it is a world, the world he inhabits, not *the* world or indeed many peoples' world. In his flattened world of meetings with important people in well-serviced globalizing cities, he sees little wrong with globalization. But his flat world is not inhabited by everyone; only a lucky few.

Yet to read Friedman is to read of the benefits of globalization for the wealthy as if they extended to everyone. Even reviewers tend to read off this privileged position as somehow standing in for something larger than the articulated experience of a well-connected, affluent, and independently wealthy white US male. Ian Parker's (2008) illuminating biographical profile of Thomas Friedman suggests that the journalist, with his rosy optimism of globalization, stands in for America itself. After reading the article, most of Friedman's columns and all of his books, I agree, but would add, only for a particular type of American. With his support of the invasion of Iraq, he is the type that sees a pre-emptive strike against a sovereign nation as justifiable. That there are many vile yet oil-less regimes round the world that the US chooses not to invade does little for its reputation and is more suggestive of an imperial power than a force for democracy. Friedman seems unable to realize that "Giving birth to a different politics" rarely comes from the muzzle of a gun, even one wielded by the US. He is also the type that from his privileged position sees little wrong with globalization. The green principles of his 2008 book, *Hot Flat and Crowded* notwithstanding, he continues to impress the planet with his intolerably heavy carbon footprint by living in the suburban McMansion, the frequent jet and car travel, the play on the chemically manicured golf courses. Parker makes much of his optimism, yet the obvious response is, given his privileged lifestyle at the apex of the globally wealthy, who wouldn't? Yes, a stand in for America, but not the best or the most representative.

We can make a distinction between flat space and smooth space. Friedman's image of globalization is of a flattened space where movement is simple and easy for the global elites; Deleuze and Guattari (1987) identify what they term smooth space, in contrast to striated space that is marked by hierarchies and rule-intensive interactions. They collapse other dualisms onto this distinction: smooth space is associated with the nomadic compared to the sedentary, full of potentiality compared with the regulations of the striated. Friedman's flat space is on closer inspection more like the striated space of Deleuze and Guattari. The emancipatory promise of globalization is of a smooth space but the current reality is a striated space.

Globalization has different effects on capital and labor and on different sectors of labor. While capital is freer to move across the surface of the globe, labor is much more restricted. The low cost of international transport and the growing ease of international trade, crucial requirements of economic globalization, have allowed capital to be more easily disassociated from national interests and local community concerns. Capital is free to roam the space of the world in search of lower wages and higher returns. Capital is now hypermobile while labor is more immobile. Workers in one factory cannot bargain in the same effective way that the workers of the old Fordist system could. Capital is no longer so fixed in place, it roams across space. Retailers can move their production contracts to another factory in another country. Organized labor is more fixed in place. The result is an uneven bargaining arrangement. Globalization has liberated capital from territory, citizens, and communities.

While capital flows more freely, the international movement of labor is tightly controlled. There are global talent pools for many economic activities, whether it be soccer players for the English Premier League, computer experts for Silicon Valley, or fashion models for designers in Milan and New York. Yet, in general, globalization now signals the power of capital to move at will, while those without capital are stuck in place. Space and place; freedom and constraint. Globalization has created large middle classes in many developing countries, but it has also given more power to the powerful and further constrained the weak. Globalization strengthens capital as well as very selected labor groups who can compete on a global level, but for many, restrictions remain in place.

There are different levels in the flows of globalization that lead to different geographies of globalization. At the top level of the flat world, capital flows with fewer and fewer restrictions. Then there are the global rich and the highly skilled, especially in the business of global business. At the bottom are the unskilled and those without capital whose movements are much more circumscribed: they live in a more rugged topography.

And cities? They are part of the peaks and the troughs of this more rugged world. In terms of individual cities we can identify three distinct types. First, there is the hierarchy of command centers: at the top are the hyperglobalized nodes, the information-gathering and processing centers at the peaks of the economic landscape. These include the Londons and New Yorks as well as the smaller national equivalents of the Sydneys and Aucklands. In the last twenty years some cities have moved higher up the ranking as they shift further up the hierarchy from globalizing to globalized; the Barcelonas and the Shanghais, Singapores and Miamis for example. Then there are the still globalizing, the Accras and the Athenses. Second, there are the production centers, the outer regions of giant metro areas around the world, where goods are made. We can identify a movement in the last half-century from the Detroits and Clevelands through the Tokyos and Singapores and Seouls to the Shanghais and Ho Chi Minh cities. The pull of cheap labor and the agglomeration effects of expanding industrial districts guides this global shift in manufacturing. Then there are the cities in the valleys of global capitalism. These black holes, as I have termed them (Short 2004b), include the Kinshahas and Pyonyangs, where even cheap labor fails to make up for all the other drawbacks of dysfunctional societies, failed states and quixotic regimes.

In terms of individual cities and citizens there is also a similar tripartite structure. There are the hyperglobalized business districts housing transnational corporations with their accompanying upmarket neighborhoods and expensive restaurants. These areas house the masters of space. At the other end are the poor residential areas where people are the prisoners of place, the ethnic ghettos of US cities, the immigrant neighborhoods of Paris, or the slum dwellers that fringe the cities of the developing world. Even the richest cities have poor areas, and even in the poorest cities there are enclaves of affluence and privilege. The inequalities are national and urban, global and local. Between the masters of space and the prisoners of place, is the precarious position of the once solid, now increasingly penumbral, middle, encompassing the upwardly mobile as well as the downwardly spiraling.

Political globalization: a neoliberal world?

Political globalization has a number of different meanings. At its simplest, it assumes a process of increasing international cooperation. Moving further along this trend line, there are some who argue that globalization heralds the demise of the nation-state. I argue that we are very far from the death of the nation-state. The nation-state is resilient, durable, and strong. Rather than globalization replacing the nation-state, there is a more complex dialectic. Political globalization is not undermining the nation-state: it is reshaping it, and globalization in turn is being channeled and shaped by nation-states, especially the more powerful states—the "great powers" as they used to be called.

The Late Modern Wave has three important dimensions of political globalization:

- the increasing number and depth of transnational organizations that enmesh nation-states in networks of global regulation;

- the resilience of the nation-state;
- the rise and resistances to neoliberalism as an ideology of both the nation-state and the global system.

First, as an exemplar of a transnational organization we can consider the case of the United Nations (UN). The UN is an important element of political globalization in the current wave of globalization. Originating in 1945, it was structured with a grudging and rationed commitment to an international organization undermining national sovereignty, especially the sovereignty of the great powers. This was evident in the creation of the two-tier system of the General Assembly and the Security Council. Every country has one vote in the General Assembly, which is a place for general discussion, but in the Security Council, the body responsible for maintaining international peace and security, there are five permanent members: China, France, the UK, Russia, and the US, as well as ten rotating members that serve for two years. Five rotate off every year, giving even more power to the permanent members. This two-tier system codifies and reinforces the hierarchy of state power. Despite the worst fears of conspiracy theorists, the UN is not heading toward a one-world government. Nations, and especially the more powerful nations, resist encroachments on national sovereignty. The principle of national sovereignty is built into the charter of the UN. No state wants to give up its national sovereignty. However, both large and small powers want some form of the UN. For the smaller powers, it gives them a voice and the possibility of influencing global debates. For the larger powers it provides a sense of global order and control: it can be used to justify and support their actions and delegitimize what they consider are rogue states.

If a world government or governance is not yet reality, a sense of global connectivity certainly is. International agreements, global discourses and the increasing internationalization of law have manufactured a global community. The UN plays an important role in the creation and maintenance of a number of discourses that, at the very least, provide the beginnings of thinking, speaking, and acting on global terms: these include the process of decolonization, the idea of universal human rights, and certain environmental issues and metrics of economic development.

An important dimension of political globalization is the continued growth of international regulatory systems and regimes. Global regulation is most evident in the trend toward economic integration. So, let us consider this in some detail by looking at the International Monetary Fund (IMF), World Bank, and World Trade Organization (WTO). Again, the roots lie in the past, most particularly in the aftermath of World War II, when the great powers were drawing up a set of rules and regulations for a new world order open to the benefits of free trade. In 1944 discussions took place at Bretton Woods, an attractive small town in New Hampshire, between the delegates from forty-four nations. John Maynard Keynes was one of Britain's representatives and one of the main architects of the plan. The delegates agreed to establish the International Monetary Fund (IMF) and the World Bank. The IMF was initially set up to provide monetary and currency stability so as to enhance world trade. Members were expected to declare fixed interest rates. The background was the severe devaluations in the interwar period, the time of people in Germany trundling wheelbarrows full of money to buy bread. Because fixed exchange rates could lead to balance of payments problems— one country could be paying more for imports than getting for exports, for example—the IMF was empowered to provide credit to member countries. The IMF was a bank that lent money to sustain the fixed exchange rate system. However, fixed exchange systems ran into difficulties. The first devaluation of the US dollar in 1971, in the wake of massive US government spending on social programs and the Vietnam War, heralded a move toward floating rates of exchange. Many countries allowed their currencies to float on the market; the value based not on some fixed figure but on what the market was willing to pay for it. The IMF would seem to be obsolete, but by the late 1970s, it reinvented itself to act more in terms of a surveillance of a country's economic policies as well as a lender of last resort. The change in the IMF embodied the changes in other international financial regulatory institutions toward a more explicitly neoliberal agenda.

The IMF currently has 186 members. Each member country contributes a certain sum of money as a credit deposit. Collectively these deposits provide a pool of money from which members may draw. The IMF acts like an international credit union for states. A country's deposits determine how much they can draw upon, known as "special drawing rights," and their voting power. The richer the country, the more it can deposit, the more it can withdraw and the more power it wields. The largest single contribution is from the US, which guarantees it around 16.7 percent of the total votes. The next largest is Japan with just over 6 percent, and then Germany with 5.8 percent, of the votes. The rich countries dominate the IMF as they contribute most of the funds and set the agenda through their voting power.

The IMF wields its power through its surveillance system and the strings it attaches to lending. Surveillance, the term as used by the IMF, involves monitoring of a member's economic policies and evaluation of their economy. IMF surveillance has been strengthened since the Mexican and Asian crises of the 1990s. The IMF evaluates countries, less in terms of their domestic policies and how they treat their citizens, and more in terms of them as functioning cogs in the global economic system.

IMF lending comes with changes in domestic policies. Borrowing countries must undertake reforms that eradicate economic difficulties and "prepare the ground for high quality economic growth." The reforms are generally devoted to reducing government expenditure, tightening monetary policy, and the privatization of public enterprises. The effects of IMF reforms are obvious. In the short to medium term, they lead to increasing unemployment, reduced social welfare spending and declining living standards for medium- to low-income groups. IMF discipline is felt most by the weakest groups in a society. The IMF operates to stifle innovative social welfare programs and redistributive policies. IMF policies tend to work against social justice.

One example: South Korea asked for IMF assistance on November 21, 1997 in the wake of depleted foreign reserves and rising short-term loans. The IMF agreed to lend $56 billion and in exchange asked for an economic adjustment package that depressed economic growth, caused unemployment, cut government spending, led to a decline in living standards, and allowed more access for foreign financial institutions. The IMF imposed market principles and freer international trade on Korean society. The IMF's argument was that South Korea's problems were due to lack of transparency in financial dealings, a bloated public sector, and a job market that ensured jobs rather than productivity.

IMF lending has been subject to two sorts of criticisms. The first has been concentrated on their regressive consequences. IMF lending requirements punish the weak and hurt the poor. A second source of criticism has been their ineffectiveness. IMF lending tends to underwrite bad investment decisions. IMF lending, in effect, pays for the poor choices of rich elites in third world countries. While investors' returns are assured, there seems little ability of the IMF to structurally readjust the economic arrangements. The first criticism says the IMF hurts the most vulnerable, the second notes that it does not penalize the corrupt elite.

The IMF is a useful scapegoat used by national governments making unpopular decisions. The South Korean government would probably have had to make structural readjustment without the IMF. With the IMF, criticism was displaced away from the national government toward the foreign other in an outburst of xenophobia. The citizens of South Korea were happy with globalization when it was assuring economic growth and generating rising employment and living standards. That was the result of Korean ingenuity and hard work. But when the globalization rollercoaster took a downward dive, now it was the foreigners and the IMF in particular to blame. Globalization is embraced when you are doing well, but criticized when things go badly.

The World Bank was also established at Bretton Woods in 1944. Its initial name, the International Bank for Reconstruction and Development (IBRD) describes its role. It was a fund established to aid the reconstruction of Europe and its first loan was $250 million to a war-damaged France. Once Europe got back on its feet, the Bank widened its remit. Until 1989 it was part of the bipolar world with funds primarily allocated to increase development to countries allied to the US. Funds went to big capital-intensive programs; the first loan to a developing country was $13.5 million given to Chile in 1948 for a hydroelectric scheme. In the early 1950s, Japan was a major recipient of World Bank loans. The loans in the early years follow a pattern of lending to US allies for large set piece capital investment programs, especially for hydroelectric power and transport developments. There was little monitoring or accountability. Much of the funding was a failure even in its own limited terms of raising aggregate economic growth.

The Bank now has five distinct parts. The IBRD provides market-based loans to middle income countries. The International Development Assistance established in 1960 gives interest free loans to low income countries; the International Finance Corporation, set up in 1956 provides loans to private investors setting up business ventures in developing countries, and the Multilateral Investment Guarantee Agency underwrites private investment in developing countries. There is also an international center for the settlement of investment disputes.

The World Bank now has almost universal membership. Like the IMF it was and still is dominated by the rich countries. By longstanding agreement, the head of the World Bank is an American while the head of the IMF is a European. Like the IMF, its headquarters is in Washington, DC, just a few blocks away from the White House. With little oversight or accountability, the World Bank became known as inefficient and an aid to rich investors and corrupt elites. In response to criticism, both internal and external, from both the developing and developed world, from both private investors angry at delays and administrative snafus, and aid workers upset at the corruption and inefficiency, the World Bank in 1989 shifted its emphasis to eradicating world poverty with more carefully assessed and monitored development schemes. It still acts in closed chambers with little input from the outside, a self-referential system that has come under increasing attack as an elite club dictating regulations in secret meetings. However, the focus of the Bank has changed toward a more accountable, local-scale, global development agency. And through its work and publications, especially the annual *World Development Report*, first published in 1978, the Bank has helped to create a global community.

The World Trade Organization (WTO) is the latest version of what was originally called the General Agreement on Tariffs and Trade (GATT) established in 1948 as a forum for stimulating world trade by reducing customs duties and lowering trade barriers. Like the IMF and World Bank it was established with an eye to avoiding the mistakes of the interwar era when rising protectionism was part cause of the Great Depression. GATT was a forum for getting rid of protectionism. The first round of multilateral discussions, held between 1948 and 1967, led to some tariff reductions. The long, drawn-out nature of the discussions led many commentators to suggest that the acronym stood for General Agreement to Talk and Talk. The most decisive round was the Uruguay round, 1986 to 1994, which established new rules for trade in services and intellectual property, new forms of dispute settlement, and trade policy reviews as well as the creation of the WTO in 1995. The Doha round began in 2001 and is not yet concluded. It deals with agricultural subsidies. The WTO now has 153 members that together account for 97 percent of the world's trade. China became a full member in 2000. The first ministerial conference took place in 1996 in Singapore, the second in 1998 in Geneva and the third in 1999 in Seattle. The seventh took place in Geneva in 2009. The WTO acts as a form of world trade court. Cases are brought to the WTO by member countries complaining of unfair trade practices. The WTO, like the IMF, scrutinizes the trade policy of the individual members.

Like the IMF, the WTO has a neoliberal agenda that promotes free trade and limited protectionism. While all members agree to this in principle, it has been difficult to work through in practice. Free trade is a great idea in principle in the long run. But in the long term, as the architect of global regulation John Maynard Keynes noted, we are all dead. It is in the short to medium term that we lead our lives. Some economists argue that only a free trade in goods and services around the world will ensure economic growth. In the short to medium term, and that is the dominant terms for most of us this side of political reality, governments have to protect the investments and jobs of domestic industries. Free trade sounds desirable in the international forum but try telling that to constituents who are about to lose their jobs or investors about to lose their investments. Every country has certain protected industries, in many cases it is agriculture, whose protection is often a necessary prerequisite to political parties wishing to retain or achieve power. This is the heart of the difficulty in reaching a conclusion to the Doha negotiations.

Most countries want to be in the WTO; they all want the benefits of world trade, but each has its own national interests. The biggest structural cleavage is between the rich and poor countries. Many poorer countries want to protect their fledgling industries from the more efficient competition from overseas. They want access to global markets for their goods but fear that their economies could be assaulted with easy entry of foreign competition. The Seattle ministerial round was broken up not only by demonstrators but also by the delegates from developing countries who refused to go along with the agreements worked out by the richer countries.

Membership of the WTO is vital to countries seeking access to global markets. But this access comes at a price. For the stronger economies free trade can mean access to new markets especially for banking and cultural economies, but it can also mean the decline of traditional industries. For the weaker economies, membership of the WTO signifies access to the large consumer markets of the rich world, but can mean an inability to control the fate of traditionally protected industries. Membership of the WTO implies a willingness to shape national economic policy around the principle of free global trade.

From the very brief review of only a few transnational institutions and regimes, it is clear that while globalization does not equal the death of the nation-state, it does lead to the growing porosity of the state. Nation-states have always existed as an international phenomenon: they are only legitimized if other states recognize them; they have borders with other countries; they trade with other countries. Today this internationalism endures, but Late Modern Wave nation-states now operate within a system of global regulation.

Their decision-making contexts and actions are permeated with global systems of control, regulation, and surveillance. Domestic policies, especially for the smaller and weaker states, are now more clearly hedged in by international regulations. The independent power of the state to operate outside these limits is severely constrained. In order to benefit from global trade, you need to be part of the global trade regulatory system, which may impose limits and boundaries to domestic policymaking.

The dominant form of political globalization is biased toward a particular purpose, the creation of more open markets. Free trade, floating exchange rates, open economies, and global competition are all aspects of globalization, and the functions of the state that implement these strategies are strengthened and reinforced.

It is not the case that the nation-state as a whole is undermined by globalization. While rapid capital flows do place limitations on the independent power of the state to influence economic forces, the state still retains immense power in regulating flows of people. While capital is less restricted by national boundaries, labor is much more controlled; the regulation of immigration is a key function of the state, and the profile of this role is currently elevated. The power of the state to limit flows of people across the national boundaries, a fundamental element of national sovereignty, is being increased partly as a response to the xenophobic fears of globalization and partly as a fear of cheaper labor undermining wages and conditions of citizens. The distinction between aliens and guests, visitors and immigrants, citizens and non-citizens is highlighted.

Globalization does raise issues of political scale. Throughout most of the nineteenth and twentieth centuries the nation-state was the scale for emancipation movements. It was "national liberation movements," "National Socialism," always *national*. The global economy has weakened the power of the individual state. Some states are less powerful than others; the US can influence global economic trends more easily than Botswana, but even the largest states react to rather than create global flows of capital. A global capitalism knows no national boundaries.

The very idea of the nation-state that is embodied in the fear of the globalization argument assumes a unified territory. A distinction can be drawn between a nation and a state. A nation is a community of people with a common identity, shared cultural values, and an attachment to a particular territory. The state is a political organization covering a particular territory. The nation is a group of people; the state is a territorial unit of political organization. There is no simple relationship between nation and state. There are nations without states, such as the Kurds, Palestinians, and the Basques. Many states have more than one nation. Even a long established democracy such as Britain contains Scots, Welsh, and Irish. While globalization may be undermining the nation-state it may aid nations against states. Nations without states may have no forum apart from international organizations. Nations being repressed by their states can appeal to international standards of human rights.

Globalization may be undermining the power of states, but it can help in the defense of nations. The Kurds continue to survive precariously in northern Iraq because of international forces. Around the world indigenous peoples ignite international public opinion to counter national indifference. Globalization can help nations survive their states. There are complex interactions between the global, national, and local that the simple thesis of "globalization undermining the nation-state" fails to comprehend. As an example that condenses a more nuanced model of global–national–local connections, in 1972 at the United Nations Educational, Scientific and Cultural Organization (UNESCO) general conference in Paris, a number of delegates addressed the problem of threats to the world's cultural and natural heritage. Noting that "the deterioration or disappearance of any item of the cultural and natural heritage constitutes a harmful impoverishment of the heritage of all nations of the world," the delegates created the World Heritage Convention (WHC), which established a list of sites to be protected. These included Stonehenge (UK), Galapagos Islands (Ecuador), Machu Picchu (Peru), and Auschwitz (Poland). By 2010 there were more than 680 cultural sites, 176 natural and twenty-five mixed sites in 148 states. Many of them protect the urban fabric. Listed sites include the walled city of Baku in Azerbaijan, the old city of Mostar in Bosnia-Herzegovina, and the historic town of Ouro Preto in Brazil. The list and the process of listing have created a global discourse of cultural and environmental preservation. The WHC has limited power, but it foments global debate that captures the local, the global and the national. Take the case of Kakadu National Park, which in 1981 became a World Heritage Area. Kakadu is home to the Mirrar people, as well as the site of one of the richest uranium deposits in the world. The federal Australian government's decision to allow uranium mining at Jabiluka, close to Kakadu, aroused the resistance of environmental groups as well as most of the Mirrar people. This coalition of green and indigenous movements lobbied the WHC to get Kakadu listed on their danger list as one way to raise international awareness of an "internationally" important site and to bring international pressure on the Australian government.

The second dimension then of political globalization in the Late Modern Wave is the continuing resilience of the nation-state. The nation-state is not dead. It is alive and well and integrated into acceptances of and resistances to globalization. Some worry that globalization has or will undermine national sovereignty. The obvious retort is, is that such a problem? Those who worry, tend to posit a more benign nation-state against an uncaring globalism. This is not always the case. In some cases nation-states oppress their populations. The UN interventions in East Timor or Kosovo were cases where states were terrorizing local populations. In these cases there was an effective and more humane alliance of the local and the global against the nastily national. Simply posing the global against the national, as the death-of-nation-states theorists tend to do, is to miss this tripartite scale in which the state does bad things against local communities whose only defense is to mobilize the global.

The Late Modern Wave of Globalization involves a shift from a state-centric to a more complex system of states with different levels of power bound by varying degrees of global regulatory networks. One example is the case of individual rights that have shifted from being the monopoly of the relationship between the state and its citizens to a more complex set of relationships among individuals, states, and global norms. Universal human rights are now often posited against the rights afforded by or transgressed by individual states. Current forms of globalization are thus generating new forms of a world polity rather than a world government.

The third dimension of political globalization is the rise and resistance to neoliberalism. Neoliberalism has emerged as the dominant ideology of the Late Modern Wave of Globalization. It began as a reaction to the rise of the Keynesian/New Deal State that was in turn a political response to the Great Depression. Faced with massively high unemployment and the real threat of political instability, economic theorists such as John Maynard Keynes and political operatives such as Franklin D. Roosevelt fashioned, almost on the fly, a model of an enlarged state responsible for mopping up unemployment, reducing business downturns, and providing a welfare net so that market downturns did not have such devastating social outcomes. The period from 1933 through the 1970s marks the high point of the Keynesian/New Deal when there was a consensus, in much of the developed world, between capital and labor on the role of government. Government spending stimulated demand so that unemployment would be limited and controlled. The social consequences of business downturns were softened through government spending on programs that ensured that the majority of the population had access to relatively affordable health, housing, education, and social welfare. In the US, business interests held a stronger hand in comparison to North West Europe where organized labor was relatively stronger and social welfare programs were not so curtailed by the greater resistance to taxes and to the role of government in general. Military spending sustained the US's global reach but restricted social welfare spending. On both sides of the North Atlantic, however, the economic muscle of organized labor impacted the deal by forcing concessions out of business and government. Life in the Keynesian/New Deal societies softened the rough edges of a capitalist economy.

There were critics of this dominant discourse. Friedrich Hayek (1899–1992) was a longstanding critic who, in his 1944 book, *The Road to Serfdom*, argued against centralized government planning. However, his was a lonely voice in the wilderness, his ideas only resuscitated in the 1970s and 1980s, quoted in praise by Margaret Thatcher and Ronald Reagan, part of the growing revival in the work of classical liberal economists—those who saw the market rather than governments as the solution to social ills. Milton Friedman (1912–2006), a Nobel-prize winning economist, also lauded the virtues of free markets and small limited governments. His book, *Free to Choose*, published in 1980, linked individual freedoms with functioning markets and restrained governments. In this restated neoliberalism, unregulated markets were the solution and governments were the problem.

Why did right-wing and then even centrist governments so warmly adopt the ideas of these post-Keynesian political and economic theorists? From the 1970s the Keynesian/New Deal consensus was under threat and then began to disappear. There were many factors at work that have been widely documented: the persistence of stagflation that seemed to disrupt the balancing act of government to minimize unemployment while avoiding inflation; growing resistance to government taxation as programs were funded by deepening and widening the income tax and local property tax base. There was also the declining power of organized labor. The loss of manufacturing jobs and the consequent decline in the size and importance of organized labor meant that business interests strengthened in comparison to a weakened labor. Beginning in the 1970s, a new metanarrative is established that limits government spending, especially on welfare programs, reduces social subsidies, deregulates markets, globalizes economies, imposes limits on tax increases, all resulting in a massive redirection of government spending and a dramatic reorientation in the role of government.

The ideas of the entrepreneurial society contested and finally dominated the Keynesian/New Deal consensus. The principles and practices of the Keynesian/New Deal compact were abandoned in favor of a neoliberal agenda of privatization, a reduced commitment of government to the welfare of its ordinary citizens and an emphasis on enhancing corporate profitability and improving business competitiveness. The global shift in manufacturing reduced the power of labor, which in turn strengthened the hand of business in shaping government policy to meet its economic agenda.

The neoliberal agenda is promoted by global systems of governance. Both the IMF and WTO, for example, promote neoliberal economic policies. Although the term is widely used and frequently cited as a coherent argument, neoliberalism is at times an inchoate grouping of ideas that coalesce around core propositions that deregulated markets and freer global trade will increase economic growth and raise living standards. Around this central ideological core there are subsidiary ideas about accountability, choice competition, incentives, and performance. Neoliberalism is also a political process that recasts citizens as consumers, reimagines states less as providers of public services and more as promoters of private growth, and shifts governance issues from citizen entitlements to consumer choices. In a full-blown neoliberalism, markets trump the state, capital wins out over organized labor, consumers replace citizens, and market choice replaces citizen rights. The private market and the accompanying individual consumerism are enthroned as the means and measure of success.

Like all economic ideologies, neoliberalism contains inconsistencies and paradoxes. The implicit assumption that freer markets lead to free societies is routinely proved as a fallacy. There is no one-to-one relationship between deregulating markets and creating political freedoms. The giant counter example to neoliberalism is the China Model. Since the 1980s China has embarked on a successful path of economic growth in which private markets are encouraged yet the central state remains large, significant, and ultimately all powerful. The impressive success of the China Model, in which over 500 million were lifted from poverty in less than thirty years, shows that economic growth can be achieved and living standards improved with a large and active central government. In 1990, 60 percent of China's population lived in extreme poverty; by 2005 the figure plummeted to 16 percent. The greatest success story in the Late Modern Wave of Globalization is not neoliberalism but the China Model.

Neoliberalism as a political theory has the benefits of being simple, easily understood and even easier to articulate. Free markets, free people. Deregulated markets generate growth. The phrases slide off the tongues of even the dimmest politicians. As a political practice, however, its greatest difficulty lies in implementation when it comes up against the brute realities of mature markets dominated by powerful sectional interests. These interests promote free trade when it meets their needs but not when it undercuts their power and market share. The tenets of neoliberalism are breached regularly in the face of political realities. Elaborate subsidies and tariff barriers, for example, maintain agricultural producers in the developed world. While efficient Japanese carmakers want free trade, inefficient Japanese rice farmers do not. The US financial services sector wants to penetrate foreign markets, yet the US agricultural lobby funds the erection of import blocking tariffs.

There are also resistances to neoliberalism. Political globalization creates spaces for global public spheres that provide sites of resistance and narratives of dissent. There has been the development of a global discourse of human rights, environmental protection, and shared projects of poverty reduction. An alternative global vision to neoliberalism is emerging. There is an increasing sense that we live in one world with shared concerns and interests. The literal image of one world is a recent phenomenon. Satellite pictures of the globe were only first published in 1969. Technology gave us, for the first time, a photographic image of one world, a shared planet. Soon after, in April 1970, Earth Day was established as a worldwide festival of one-world environmentalism.

The means of communication also shapes our view of the world. When we are limited by word of mouth, our horizons are limited to the very local. But with the evolution of print, messages can be carried further, communities of readers are extended across space and time. With electronic media, communities of viewers and listeners can stretch around the world, creating a global community. The global is brought into our living rooms. Events happening around the world are viewed as they happen: indeed they can happen *because* they are on television. In 1964 Marshall McLuhan referred to the global village. I see it more as an *occasional* global village, only certain events are dramatized and selectively broadcast. There is a new force in world affairs: global public opinion that can be mobilized to release prisoners, undermine governments, legitimize opposition movements, and send aid to starving people. There is always the danger of compassion fatigue, as yet more images of starving babies are easier to ignore. But global public opinion, fickle and changeable, ephemeral and unfocused, is now a fact of life. It is one of the more positive aspects of political globalization. When East Germans demonstrated in Leipzig in 1989 against communist rule, their resistance was seen all over the world, and it gave hope and confidence to dissidents in Czechoslovakia and Romania. It is unlikely Nelson Mandela's release would ever have taken place without the unblinking gaze of world opinion. The image of the global, the creation of global public opinion through mass media, the time–space compression of telecommunications and transport improvements that brings the world much closer together, have all created the preconditions of global discourse. However, although global images flit occasionally past our eyes, the everyday representation of our lives is still fundamentally shaped by our national location. Who we are is still very much a function of where we are. We read national press, watch national television: the education system is in large part an ideological exercise in national consciousness. One of the most important documents when traveling the world is our passport, which denotes our citizenship and enables our international mobility. We are still very much stamped by the imprint of the national. A global consciousness develops only under the heavy impress of the national.

There is the global consciousness of the transnational business class, the international bureaucrat or the conference-hopping intellectuals who know a good tailor in Hong Kong, the best hotel in Kuala Lumpur, or where to find the best bagels in New York. These jet-setting, business-class traveling sophisticates' global perspective is from the concierge level in a luxury hotel, a perspective unlikely to foster emancipatory projects for the masses. Generating global citizenship on a mass scale for emancipatory social projects is difficult. While private capital flows freely and unelected regulatory bodies such as the IMF and WTO set global rules, our citizenship of the world is given few ways to express itself. We only indirectly vote for the people who represent us at the United Nations. So far, private and bureaucratic interests have dominated a globalization project that so far is not subjected to democratic accountability.

There are sites of resistance. The WTO held a ministerial meeting in Seattle in November–December 1999 to launch a new round of trade liberalization talks. Over 135 countries were represented. The event seemed to galvanize a wide body of opinion. Almost 1,200 non-government organizations in eighty-seven countries signed a petition calling for a fundamental reform of the WTO. Groups as varied as the Self Development of Indigenous People in Mexico and the Friends of the Earth joined in common cause to condemn the present system of world trade and the closed and regressive nature of WTO rules and regulations. Such groups as the Direct Action Network organized effective public protest. On Tuesday November 30 the official opening of the conference was delayed as almost 20,000 people blocked the delegates entering the meeting at the Washington State Convention and Trade Center and effectively closed down the center of the city. Over the next few days, protest actions were seen around the world and the WTO, which had been an organization unknown to the wider public, was placed before the harsh glare of media attention. The next day estimates of the crowds had increased to 50,000. While images of property destruction were widely broadcast—the destruction of a Starbucks coffee house was given coverage in many newspaper reports—the demonstrations succeeded in shifting public opinion toward a closer scrutiny of the WTO and the supposed benefits of free trade. The complaints widened to a critique of consumerist society, declining environmental quality and the power of multinationals. The WTO was put on the defensive. The talks were essentially canceled not only because of the protest, but also because the delegates could not sign off on an agreement. Many delegates from developing countries were buttressed in their concerns by the 50,000 protestors.

The success of protest in Seattle led to plans for protests at the joint IMF/World Bank meeting in Washington, April 16–17, 2000. Estimates of the crowd varied from 35,000 according to the organizers and 10,000 according to the police. Whatever the numbers, the organized crowds were successful in closing down the city and drawing public attention to the two organizations. The meetings went ahead, and the police, learning from the Seattle experience, successfully managed to get delegates to the meetings. More than 1,500 officers in riot gear handled the protestors with varying degrees of consideration. However, the institutions responded to the criticism. Although he disagreed with their methods, one delegate, the Argentine treasury secretary, noted, "I think the protestors perhaps have some points" (quoted in Finnegan 2000, p. 42). On 16 April, the IMF issued a communiqué that admitted the growing discontent with globalization and that prosperity was not reaching everyone. The IMF announced initiatives for renewed emphasis on debt relief for poorer nations and tighter auditing of IMF loans to avoid abuse. The next day the World Bank issued a statement that pledged to give more money to fight AIDS, and speed up debt relief for developing nations. The president of the Bank, James Wolfensohn, claimed that the Bank was focusing its energies on small-scale programs aimed at eradicating poverty.

Political globalization is primarily concerned, so far, with the organization of rules that emphasize free trade, deregulation, and privatization, and favor corporate capital rather than labor interests and the environment. However, there are the beginnings of a global civil society that is as much international and cosmopolitan as national and parochial. It is difficult to sustain. Our strongest communities tend to be those where face-to-face contact is maintained at regular intervals in sustained connections. But the Internet, cheaper travel and above all a shared sense of global citizenry is slowly emerging. So far it is most obvious in the case of protest. For most people globalization is endured and experienced, not controlled or managed. The internationalization of political discourse is only just emerging as a subject of citizen involvement and engagement.

When asked where he came from, Diogenes famously said, "I am a citizen of the world." The Stoics argued that each of us lives in two communities, the local community of our birth, and the broader global community of human reason and aspiration. Global citizenship was encouraged to reinforce and strengthen local connections, not instead of them. The pulls of patriotism are many compared to the fragmentary, ill-defined benefits of cosmopolitanism. And yet, when we recognize our moral commitments to the global community and extend the scope of national community to international community then, and only then, will a real and meaningful political globalization take place. It is only with an international civil society, engaged global citizens, and a global moral consciousness that political globalization can live up to its emancipatory promise.

The greatest criticism of neoliberalism emerges from political realities. Just as the Great Depression highlighted the inadequacies of market solutions and produced the Keynesian/New Deal consensus so the Great Recession of 2009 that started in 2008 revealed the limits of neoliberalism. On October 9, 2007 the Dow Jones Industrial Average was at 14,165. By March 9, 2009 it fell over 53 percent to 6,678. The stock of companies previously considered blue chip fared even more dramatically worse: Citibank lost over 95 percent of its stock value in little more than a year. As Marx and Engels first noted, "all that is solid melts into air." But it is not just the value of companies and stock markets that is plummeting; so is the intellectual stock of ideas that underpins the present crisis. Three ideas in particular are worthy of some note.

The first is the idea of small government. In his first inaugural address delivered on the west front of the US Capitol on January 20, 1981, Ronald Reagan said, "In this present crisis, government is not the solution to our problem; government is the problem." No American president, or indeed any government leader this side of sanity, would use the same or similar words to address the present crisis. There is now recognition that markets left to their own devices can wreak havoc as well as bring economic growth. There is a sense that government is the problem solver of last resort when markets fail to work. Just as in the Great Depression of the past, so in the Great Recession of the present, we discover again that governments are important and have a vital role in economic growth and management.

The second idea that lost value is the notion that deregulation is the cure for our ills. Markets are always embedded in cultures, societies and states, so there is really no such thing as a pure market. The call for deregulation was always really a demand for an overturning of government oversight, mandated standards and systems of control. In the wake of the economic meltdown these now seem like good things; it is as if we have rediscovered previously undervalued stocks that are holding their value in the present crisis.

The third idea is that financial globalization, especially the free flow of capital, is a good thing that needs to be encouraged by little or no controls. Unfettered, unregulated flows, especially of the more exotic credit swaps, hedge bets and futures trading have been creating major problems for years. We had a preview of their inherent instabilities. There was the case of Nick Leeson, a trader in Singapore who brought Barings Bank to collapse with not very smart trades in Japanese stock index futures. In 1995 his losses reached $1.4 billion and the bank was declared insolvent. Then there is the futures trader Jerome Kerviel who in 2008 cost the French bank Société Général $7.14 billion, about a fifth of its total capitalization. Both incidents were designated as problems of rogue traders whereas in fact they reflected, embodied, and predicted systemic failings of the entire system. We cannot say there were no warnings.

To use the language of the market: it is crystal clear that we need to re-evaluate our intellectual stock portfolio. The blue chips of small government, deregulation, and the unregulated flow of capital are failing and falling. We need to invest in some new, or in many cases old and undervalued, ideas that may give us a better return and create some longer lasting value. This recognition does not answer the question of what precise role the government should play. The original New Deal maintained monopolies, kept prices high and while it did reduce unemployment, it was always tinged with a narrow economic nationalism that needs to be avoided in the present circumstances of such a linked global system. So let's have a new term to mark a new beginning, perhaps Global Compact or, as a nod to an even older reference, The New Social Contract.

And what of cities in political globalization of this Late Modern Wave? A number of trends can be noted:

* the shift from the Keynesian/New Deal to the Entrepreneurial City;
* new regimes of urban governance;
* growing inter-city competition.

The decline of the Keynesian/New Deal consensus and the rise of neoliberalism is also apparent at the level of cities in the emergence of the Entrepreneurial City, marked by an encouragement of private growth and the creation of profit-seeking public/private partnerships. Concern for the social welfare of citizens is replaced by the need to enhance the competitiveness of local business through tax reductions and the attraction of footloose capital through attractive incentives. The city becomes a place of growth promotion as stimulating private growth trumps social redistribution, commodified spaces replace public places and citizens are recast as consumers. The dominant form of urban governance shifts from a welfare state model to an economic development model. The hand of growth coalitions is strengthened while the redistributional arguments are marginalized. To be sure, there are differences between cities as this shift passes through the filtering membranes of national and local cultures. But, in general, social programs are subordinated to business interests, which are institutionalized into the emerging metanarrative of urban political discourse.

There is now a considerable literature on what is variously described as the urban growth machine, city boosterism, and urban entrepreneurialism, the process whereby city elites promote the economic competitiveness of the city through attracting investment and spurring economic growth. John Logan and Harvey Molotch (1987) identify an urban growth machine formed by powerful economic interests that seek economic growth as the city's main political agenda. David Harvey (1989a) also points to the shift in urban governance from an urban managerialism to an urban entrepreneurialism. More recent work has refined these early studies (Hall and Hubbard 1998; Jonas and Wilson 1999). There is also more detailed work on the explicit use of imagery in the selling of the city (Short and Kim 1998; Short 1999; Kim and Short 2008).

Most city policy agendas combine elements of both economic growth and social development into overlapping programs for workforce development, job creation, and community development. Economic development receives the highest priority from politicians in large cities in the developed world. Politicians often build campaigns around the positive message of economic growth and of national and global competitiveness. O'Flaherty (2005) lists several prominent policy initiatives designed to promote urban economic development and, more specifically, to make a city become desirable for potential investors. They include: lowering tax rates, passing targeted tax cuts, offering tailored incentives such as subsidies through infrastructure, land or training programs linking specific incentives to specific outcomes, reducing regulations, allocating urban enterprise and empowerment zones for economically distressed inner city neighborhoods, and building new sports stadiums and arenas. Increasing national and global competitiveness is a crucial element in defining these new regimes of urban governance.

Urban politicians also aggressively promote the vision of their city as a world-class city when they solicit public support for controversial plans, such as urban redevelopment projects, large-scale infrastructural upgrading, and hosting or bidding for high-profile international events; when arguing for large-scale tax incentives and business-friendly economic policies; and when promoting attempts to heighten multiculturalism and, in a broader sense, cosmopolitanism among city residents.

There is now a dominant global city imaginary that embodies ideas of global competitiveness, neoliberal agendas, cosmopolitan cultures, and spectacular mega projects. The city is reimagined as competing on the global economic stage as active competitor full of jaw-dropping structures with a tolerant and business-friendly environment. Golubchikov (2010) provides one such case study in his discussion of the production of a new St Petersburg, a city reimagined by Russian elites as an international hub of circulatory capital, host to corporate power and stage for globalist megaprojects. The process is a marked shift from the old Soviet regime of urban subordination and social redistribution. In this recent reglobalization St Petersburg is recast in a new neoliberal regime as an entrepreneurial, economically competitive, globally connected city.

A world city imaginary that nurtures cosmopolitan cultures helps city governments solicit public support for costly cultural projects, such as ethnic festivals or the construction or renovation of cultural centers. Cosmopolitan cultures often sustain successful tourist industries—one of the target industries, along with high-tech industries, that almost all cities include in their urban economic development initiatives.

The adoption of neoliberal policies promotes and enhances competition between cities at both the global and national levels. A more deregulated market allows individual cities to break away from the confines of national regulation and the constraints of a national market. The fate of individual city economies is cut adrift from the national economy as national planning and regional policies are jettisoned in favor of more market-led approaches to urban competitiveness.

As the world flattens, there is growing competition between cities. The larger cities compete for advanced producer services while the production centers compete for manufacturing business. And as the world flattens, cities in many different countries are brought into greater international competition whether it be attracting corporate power and financial concentrations, encouraging manufacturing or promoting tourism. At all levels of the urban hierarchy there is increased global competition. Local urban politics become wrapped in the discourse of globalizing the city. Urban politics is constrained by the competitive realities of a flattened world.

In this cauldron of intense competition, cities are reimagined and represented as business-friendly, growth-enhancing places. To these business-friendly agendas is added a repertoire of other images: socially tolerant, cosmopolitan efficient and attractive, a great place to do business, a great place to live. These themes are repeated in the images and new narratives of the global city imaginary. As the world flattens and locational flexibility in many sectors increases, cities now compete one with another for business, investment, people, and capital. The work of signature architects and global events and spectacle are the new currency that spur and embody this increased urban competitiveness.

We can identify a new urban regime, *the globalizing entrepreneurial regime* that in its purest form is marked by a commitment to acquiring and maintaining global city status through a form of urban entrepreneurialism that proactively fosters local economic development to ensure global competitiveness. This regime makes full use of the market, has a strong pro-business rhetoric and practice and actively employs public–private connections to competitively position the urban economy in the competitive global marketplace.

Cultural globalization: a homogeneous world?

The term *culture*, as Raymond Williams (1976, p. 76) wrote, is "one of the two or three most complicated words in the English language." The word has complex multilayered meanings. One important notion that arose with Romanticism is that "culture" refers to folkways counterpoised to the materialism of industrial modernity. Culture was what the people did in their everyday lives. There was, however, a class distinction between low and high culture that drew attention to both the local and ostensibly global cultures: folk tunes compared to symphonic music, for example. But even here the interpenetration of the two was evident, whether it was in the resuscitation of traditional Scottish ballads by Robert Burns for a broader audience or the use of folk songs and melodies in the work of the great Romantics such as Frederic Chopin, who was both a composer of international distinction as well as a deeply nationalist composer drawing upon traditional Polish dance melodies. I will use this idea of the interpenetration, the interweaving, the folding and unfolding of the local, the national and the global components of what we mean when we say "culture."

While most agree that continuous, increasing and deepening flows of goods, people, capital, ideas and information across national boundaries culturally globalize the world, one perspective tends to identify a more homogeneous world. Those who favor the cultural homogenization account point to the formation of a singular global consumer culture. For some, this globalization is a form of American cultural imperialism as more people drink Coca Cola, eat McDonald's, and watch Disney movies. The fact that people across the globe are watching CNN and MTV, that McDonald's are opening throughout the world, and that many Hollywood films dominate the world film market, are taken as indisputable evidences of the homogenization and Americanization of the world (see Figure 2.2).

The alternative thesis, the complex interweaving of local, national and global that I subscribe to, argues that, while particular television programs, sport spectacles, network news, advertisements and films may rapidly encircle the globe, this does not mean that the responses of those viewing and listening will be uniform. Goods, ideas, and symbols may be diffused globally, but they are consumed within national and local cultures. Eating a Big Mac does not make one an apologist for US foreign policy. Ideas, symbols, and goods that circulate around the world are consumed in distinct national contexts and in unique local circumstances. Specific cultural backgrounds are not just empty containers for the receipt of global messages, they are critical to how messages are received, commodities consumed and ideas are interpreted. People in the contemporary world are increasingly familiar with the presence of different cultures, rather than sucked into a single cultural orbit.

Cultures are not static, fixed in time and place. A more subtle view of culture and cultures sees it and them as resulting from ongoing interactions between the local, the global, and the national. The processes that shape local and national cultures are not one-way interactions, but are rather dynamic

Figure 2.2 Cultural globalization in the city: golden arches in Shanghai.
Photo: John Rennie Short

and multifaceted, so that hybrids of the "newly arrived" and the "previously there" are constantly reconfigured and remobilized in and through global flows. Culture is not just passive audiences watching imported television programs or eating fast food from global chains. Culture is an active process: one definition of culture is to tend, to husband resources; the same root as cultivation. A culture is not just consuming things; it is the production of a worldview, an aesthetic sensibility, the active creation of a history and geography, the working out of a place in the world. Cultural expression is as much a fact as cultural consumption. Culture is associated with place in complex ways; it is more than just a passive receptor, the endpoint on the pathway of global commodification. Culture involves the active use of these global flows into local, national, and group specific views of the world. In fact, cultural globalization has led to a more explicit concern with more local cultures as certain groups seek to redefine or re-establish their place in the world.

When we speak about culture in the current debate on globalization, there is a tendency to counterpoise a local unchanging culture with a new, bland, global culture. But cultures are not unchanging, static remnants of bygone ages. Cultures are active processes of assimilation and hybridization. All cultures are mixtures of the global and the local, the past and the present. Cultures are not pure states; they are active processes of hybridization. Cultures appropriate and mimic other cultures and have been doing so since the beginning of time. The image of local authentic cultures being overwhelmed by global commodity culture is to ignore the mixing already embodied in local cultures.

We can replace the popular image of local cultures under assault from a global culture with a more complex picture of local cultures being generated and recreated in response to globalization. The sense of a pervasive globalization has not so much overwhelmed local cultures as helped create them. The search for authenticity, for pure local cultures (which are in fact mixtures) has helped in the creation of world music, ethnic cuisine, and indigenous culture. Religious fundamentalism is less a return to a pure theocratic ideology and more the self-conscious recreation of religious beliefs in the face of secularization and globalization. Cultural authenticity is less an excavation of the pure and more a contemporary representation of the contrived. There is no simple division between a local and a global culture. The co-presence of homogenizing and heterogenizing trends might be a better phrase to describe the processes of cultural globalization rather than a binary classification of globalizing/non-globalizing. We can unravel some of this complexity even further by looking in detail at:

- the commodification of culture
- the deterritorialization of culture.

Cultural globalization has been reinforced by the capitalization of culture. Sports, music, art, cinema, and dance have all been commodified. The commodification of culture does lead to a certain similarity in the cultural mix available to audiences around the world. But the more astute companies tailor their cultural products for specific audiences. Hybridity is a feature of globalized commodified culture. Multinational corporations such as McDonald's, Guinness, and Coca-Cola are adopting practices of hybridization to make their global staples local favorites. Guinness hires locals in its factory in Accra, Ghana, and McDonald's includes vegetarian menu items in India. One of the guiding directives of presentation for these commercial giants is that successful brands are personal, an integral part of people's lives. Engaging with local cultures and hybridizing global brands forge emotional connections. Akio Morita, the legendary head of Sony, first coined the term "global localization." It was a business strategy to create worldwide operations that were attuned to local markets and conditions. Global consumers and global companies are creating a globalized consumer industry, albeit tailored to specific national markets and local conditions. The commodification of culture is closely connected to the globalization of the economy.

When we think of culture we often think of distinctive songs, dances, foods—all those things that make up collective memories and group identities. In much of our understanding, these are all tied to place: Mexican food, Indian music, Italian design, Armenian language. But international migrations, the flows of people and ideas, the commodification and transnationalization of cultural aspects have all broken the simple connection between culture and place. On closer inspection the connection between culture and place is always complex if not problematic. "Mexican" food is a mixture of Spanish and indigenous cuisines that have been evolving in different parts of the world in different ways for almost five hundred years. The process of deterritorialization has been going on before the current round of globalization. When Spaniards came to the New World, and then Africans were shipped across the ocean, they combined with indigenous people to create a New World cuisine that was creolized and hybridized right from the beginning. There is no unchanging New World cuisine. Mexican food is itself a changing hybrid.

Cultures are always in the process of deterritorialization and reterritorialization. The process has quickened and diffused more widely, but it is not a new process. Globalization, it has been argued, undermines local identities. However, an alternative case can be made that globalization has both deterritorialized and strengthened local cultures. In the nineteenth century there was a massive immigration to the US from many European countries. Italians, Germans, Poles, and Irish all moved in their millions to the US. Their connection with home was always limited, separated as they were by rudimentary communication links. Letters could take weeks. New identities were created. People from different parts of what is now Italy became Italians in the US. New identities were shaped. Irish Americans took on the mythic elements of Irish nationalism. Folk memories did not die, indeed in many cases they became both strengthened and stuck in time. Irish Americans, unlike the Irish, could scarcely "remember" anything other than the famine. In recent years however, immigrants to the US can keep closer connections with their families back home. Cheaper telephone rates, email, more affordable plane journeys; all those easier communication systems of a globalized world have allowed groups to be more in touch with their home areas. As an immigrant to the US, for example, I can keep in touch with my extended family around the world through telephone and regular visits. I am now closer to home than I have ever been simply because of globalization. Of course what home represents is always changing.

My contention is that globalization has not disrupted the long connection between culture and place. Globalization has put new wrinkles in the connection. The more recent round of globalization has done three things: in its commodification of selected cultural forms it has transformed some "local" cultures into globalized forms. Hollywood movies for example, are becoming more international, dealing with broader themes than simply US concerns. Indian restaurants serve food that is only ever found in Indian restaurants not located in India. Second, it has also involved the glocalization of global brands into local and national markets. Third, it has rendered more complex the relationship between identity and place. The flows of ideas and images allow reterritorialization of cultures. Easier flows of people and money allow more connected diasporas and thus group identity being shaped by outsiders as well as insiders. In the Late Modern Wave of Globalization cultural globalization is creating as much difference as similarity. New cultural identities are being created around hybrid forms as well as around invented traditions.

Cities play an important role in the cultural globalization of this Late Modern Wave. I suggest they are places of reterritorializations of culture and concentrations of cultural economies. One way that the reterritorialization of culture occurs is through international migration. International migration is routed very largely through major metropolitan areas. In cities around the world the number of the foreign born has increased. Global and globalizing cities are now places of minority populations, diasporic communities and hybrid identities. The metropolis is now the place of the Other. And the Other with their distinctive languages, cuisines, and cultures now add to the cosmopolitan mix of the big cities.

International migration occurs at a variety of socio-economic levels. The Other is sometimes celebrated, as in the more cosmopolitan cities, but also curtailed in others. There are multiple migrant experiences. Take the case of Dubai where there are two very different streams of migrants. There is the encouragement of the investors, technical experts, and those associated with flight capital. Then there are the building worker migrants, often coming from rural communities in Pakistan and Bangladesh, who live many to a room in tough working conditions and difficult living conditions. Their entry is tightly controlled and monitored. Luxury homes and migrant camps, foreign investors and building laborers, welcome guests and barely tolerated foreign workers: these are the divergent migrant experiences. There are also the expatriate communities of technical workers and experts—English financial specialists in Singapore, Scottish engineers in Saudi Arabia, European aid workers in Nairobi, Indian doctors in the US; and there are the unskilled workers—the Filipino domestic servants in Hong Kong, the Mexican gardeners in California, and the vast armies of undocumented and illegal workers in shadow urban economies around the world.

Migrant communities in cities throughout world are an important network of information flows, capital flows and economic exchanges. Chains of migration are routed through families and friends, business deals are conducted among ethnic minorities spread through the world, and money is sent back from places of work to home. Economic globalization is intimately linked to cultural globalization. Businesses are culturally embedded. A flattening world is linked to the deepening interaction of far-flung communities.

Collectively these communities shape both the destination and origin areas. When migrant communities are established, they bring their culture with them. Sometimes it is frozen in the time of departure and lasts longer than it has at home. But even then cultures always have enough plasticity to adapt to the new and the foreign. Cities are places where otherness is renewed and celebrated, performed and remembered in creative acts of representation and memory. Otherness is reinforced in the large metropolitan centers because there is often enough mass for minority status to be confirmed and represented. And this representation adds to the cosmopolitan mix of the city. Where tolerance is practiced, Otherness is celebrated. Otherness is disciplined in the face of intolerance. There is undeniable global mixing in the large metropolitan centers. And there is no going back to an unchanged homeland as return flows of money and people gradually shift the local culture. Both home and away, origin and destination are constantly changing in response to the continual interactions and flows. Cheaper travel, the ease of transmitting money and ease of communications all allow diasporic communities to remain in touch, to influence their new surroundings and to be subtly transformed in the process of being diasporic.

Cosmopolitanism receives considerable attention in the philosophical literature. It is defined, drawing on the Greek words for universe and city, as a belief in shared global community with universal values. The interest reflects the strengthening global connections and the emerging global civic discourse. The starting point for many is Kant who argued for a principle of universal hospitality. Appiah (2006) is the latest in a long line of philosophers who teases out some of the dilemmas and difficulties of a practicing cosmopolitanism. However, what is lacking in these discussions is any real sense of the real world examples of the contemporary city. The discussions quickly take off into the airy world of ethics when they could be better grounded in the everyday urban experience of contemporary metropolitan regions, where difference and otherness is daily being created, negotiated, and contested. The "universal city" is not just metaphor. In globalized and globalizing cities, it is lived experience.

Cities also play an important role in the commodification of culture and the agglomeration of cultural economies. Here, I draw heavily on Kim and Short (2008). In his 1991 book *Postmodernism or The Cultural Logic of Late Capitalism*, Fredric Jameson notes that, in the latest stage of capitalism, symbolic meanings and associations increasingly determine the economic value of goods. Imposing symbolism, meanings, values, and emotions onto goods blurs the boundary between the image of those products and their concrete reality, and suggests an intrinsic link between economics and culture. The cultural industries includes music, dance, theater, literature, the visual arts, crafts, and many newer forms of practice such as video art, performance art, computer and multimedia art. Allen Scott (2000) introduces a broader list of "cultural-products sectors, " which includes high fashion, furniture design, news media, jewelry, advertising, and architecture. Sports industries, both professional sports and health and fitness-related businesses, could be added to the list.

Cultural industries concentrate in world cities and other large cities. Indeed, they contribute immensely to the economies of cities such as Los Angeles and Paris. Path-dependent theories claim that small historic events or locational advantages can affect macroeconomic consequences that privilege certain paths to development and limit others. Although criticized for its overly structural approach to local economic development, this idea sheds light on the prestigious success that Paris enjoys in high fashion, New York in advertising and Los Angeles in motion picture entertainment. Their leading roles in these industries have long enjoyed wide recognition, and their advantages over potential competitors relate not only to the quality of their products, but also to the "symbolic images, " such as authenticity and reputation, that those products carry.

Cultural concerns now play a significant role in urban development issues. A great number of inner city neighborhoods have been "gentrified" with art museums, historic districts, and, more controversially, professional sports stadiums. City governments seeking solutions for their struggling economies increasingly turn to postindustrial and postmodern economic sources. Cultural activities, previously deemed to have marginal effects on the city's overall economic health, have won a newfound respect from policymakers and, subsequently, increased public investment. City governments now consider funding cultural and arts institutions as an economic development measure. Cultural strategies of urban development have focused on events, such as ethnic festivals, cultural exhibits, performing arts, and historical re-enactments, among others. City governments expect these events to attract tourists and suburban residents to their city center, providing indirect benefits to restaurants and hotel businesses in the downtown area. While such economic benefits may or may not materialize, cities are now the stages for the presentation of culture, leisure, cosmopolitanism, and postmodernism (see Figure 2.3).

The notion of path dependence poses questions: can smaller cities cultivate local preconditions for a competitive cultural industry? Can cultural industries

Figure 2.3 Cultural economy and the city: the Guggenheim in New York City.
Photo: John Rennie Short.

bring sizable employment, tax revenues and multiplier effects to other industries in medium-sized cities, even after their governments provide them with tax breaks or other subsidies? For many cities, especially the postindustrial cities suffering economic decline and severe job loss, the cultural turn may be the only solution left.

The cultural turn in urban development projects an image of a globalizing city, one firmly linked into the global circuits of culture, taste, and aesthetic sensibilities. Hiring a big name architect, promoting the image of a tolerant cosmopolitan culture, establishing a museum and arts center that display international shows and transnational art trends are all part of the cultural capital that cities need to accumulate in order to be seen as a global contender. Cosmopolitan diversity and cultural sophistication are as much essential ingredients of successful global cities as international airports and luxury hotels. Cities capitalize a globalizing culture. Culture is commodified in globalizing cities.

———————

CASE STUDY 2.1 **Barcelona**

I love Barcelona. My love affair now stretches back over thirty-five years. I first went to the city in the 1970s. After a long train journey from London, traveling through the night across France, changing trains at the border between Spain and France, I arrived at the Barcelona Sants Station. At the foot of the Ramblas, this station was the gateway to a vibrant metropolis, so charged with energy and difference it was scary and exciting, overwhelming and stimulating. Compared to the sedate British towns of my youth, it was wondrously excessive. Neither Spanish—too much Catalan for that—nor Southern European—there was too much of a hard commercial sense for that—the city was unique, old, and modern combined in a complex urban fabric. The old Gothic Quarter, the softened-edge grid of the nineteenth-century extension and the legacy of Gaudi and Catalan modernism all combined to give Barcelona an architectural dazzle. I stayed in a cheap hotel in the old quarter where if you came back after 10 p.m. you clapped your hands in the street, and someone threw down a key. Barcelona was hot and humid and pungent. And I loved it.

I have visited Barcelona regularly but infrequently since then, and between those intervals, both the city and I have changed. I became more traveled, more used to big, strange cities. And Barcelona changed in turn: it became the capital of a more autonomous Catalan in a more democratic Spain. It hosted the Olympics; it became "cool," a fashionable destination for tourists from across Europe and around the world, now a pilgrimage destination for Japanese lovers of Gaudi. And in this process, the locals again rediscovered the wonders of their city, which is a welcome yet unexpected artifact of Barcelona's globalization.

Indeed, the city is now more cosmopolitan, more used to the foreign Other and the international tourist. While the tourist crowds overwhelm parts of the city, the city is still too big, too complex, too much still in the process of becoming, and the vibrant street life and commercial vitality too exuberant for it to become a place at rest, the mere endpoint of tourist pilgrimages. Including my own.

CASE STUDY 2.2 **Warsaw**

I visited Warsaw for one week in mid March in 2010. My first day there was wintry. On my last day it felt like spring, encouraging people to shed their heavy coats. Easy to see this experience as a metaphor for Poland's recent history: the nation emerging from almost fifty years of communist rule in 1989 to escape from Russia's icy grip. Poland embraced market reforms and joined the EU in 2004.

Warsaw's communist legacy is still evident. The Palace of Culture and Science (formerly Joseph Stalin Palace of Culture and Science) was the tallest building in Europe until 1957 and still dominates the skyline in the city center (see Figure 2.4). An exuberant example of Soviet Realism, it was completed in 1955. There are also the blocks of former public housing, modernist monoliths scattered throughout Warsaw.

Polish Communism always had to deal with Polish nationalism. The old city reflects the commitment to rebuilding the city after the Nazis nearly annihilated it. Between the end of World War II and 1962, the old town was carefully recreated from its ruins in an act of architectural memorialization that embodied Polish resilience and Polish identity. The many marvelous examples of Baroque and Neoclassical architecture remind us of Warsaw's long connection with Europe.

The command economy still lives on in an unclear and uncertain land market. Large blocks of land lie undeveloped and undercapitalized, as investors are unsure of the property rights and legal claims. But there is also tremendous growth, especially at the edges of the city as the interstices between the modernist blocks are filled in with new housing and commerce and retail. One distinct feature is the number of gated communities. Guards and gates mark off many developments in both the central city and in the suburbs.

Like many postsocialist cities, Warsaw has an infrastructure deficit, a huge housing shortage creating an overpriced housing market, a degraded urban environment in some places with feverish new investments in others. The city is still very Polish, one of the few examples of a European city that over the twentieth century has become more homogeneous. Almost a half million Jews lived in this city prior to 1939. Now there are little more than 15,000. And yet to walk along Nowy Swiat is to see Starbucks and Subway, Colombian coffee houses and Indian restaurants. As I made my way up to the old town, I could see the local fashionistas sporting the "new black" of purple scarves and coats. Cellphone usage is endemic and universal. Poland has such a poor landline network that consumers jumped more quickly to cellphone usage than in the US. EU funding projects are ubiquitous throughout the city—from refurbishing seventeenth-century palaces to providing school playgrounds.

Figure 2.4 Palace of Culture and Science, Warsaw.
Photo: John Rennie Short.

In the city and the country as a whole there is new realization that a postsocialist city does not necessarily mean a neoliberal city. Warsaw is in *Mitteleuropa*— situated between East and West Europe. It also occupies the difficult space between command and market economies, socialist and capitalist ideologies, authoritarian and democratic systems of government, a city of becoming as well as being.

3   Metropolitan modernities

The ongoing relationship between the construction of modernity and the creation of a metropolitan society is the subject of this chapter. I will argue that the relationship, always present, is now stronger and more globalized. It is now possible to identify a metropolitan modernity or, to be more precise, global metropolitan modernities. Let me begin with some definitions, not as endpoints but as apertures into wider debates.

I will use the adjective "metropolitan" to refer to big cityness. The size refers not just to the sheer weight of the urban population, although the very largest of cities do pose particular challenges and opportunities, but to the depth and breadth of a city's connections with the rest of the world. Weimar, a small town in Germany, barely registers as a metropolitan center if all we were to measure were the number of inhabitants. But this small town, as the home to Goethe and Schiller and especially to the Bauhaus—see Chapter 7 —has links with the architectural and aesthetic concepts and practices that stretched and continue to stretch across the globe, cosmopolitan then and globalizing now.

Modernity. A tricky word whose meaning is not easily expressed in a few phrases. It is one of those words nested within the general term "modern" that first emerges in the Renaissance to refer to the present, the now, and the new-fashioned compared to the ancient, the then, and the old-fashioned. Other surrounding words also can be seen in these simple terms: modernism is simply a support of modern ways, and modernity is the condition of being modern, referring to a specific time and place of new social relationships and spatial conditions. Modernity's etymology, then, refers to the here and the now, and this origin inflects the problematic temporality of globalization.

In the more academic literature the "then and before" is characterized as agrarian, feudal, with social thought dominated by a religiosity. The "now and after" of modernity is thus associated with industrialization, capitalism, and enlightened social thinking. Each of these has a distinct urban character.

Cities were the socio-spatial fulcrums of the shift from an agrarian feudalism to an industrial capitalism. Take the case of Manchester, the first industrial city. In 1774 it was just one of many textile towns in Europe with a population of just over 41,000. By 1831 it had reached 270,901 as mechanized textile manufacturing expanded. Cotton mills worked day and night, and visitors came from all over the world to marvel and criticize the modernity of industry. Even the word "industry" changed from referring to hard work to a new sector of the economy. The city grew because of developments in technology, new sources of supply and demand and social relations that fostered innovation and risk-taking. A large middle class of small capitalists utilized and created a local culture that employed entrepreneurial talent and social networks that allowed constant improvement in products and processes. Workers migrated in the thousands to the booming factories. They worked and lived in appalling conditions as new ways of working and housing workers were established with few controls or limits. New antagonistic social classes were also formed as a hell-hole of a city emerged, vividly described by Friedrich Engels in *The Condition of the English Working Class in 1844*. The industrial city was the symbol of the new capitalist industrial order and also the stage for the emergence of organized working classes, groups of workers whose proximity and shared experience created what Marx called "classes for themselves." The urban condition of early industrial capitalism, despite its searing conditions, brought workers together and saved them from what Marx and Engels described as the "idiocy of rural life"; in some English translations the phrase is rendered as the "isolation of rural life." A working-class consciousness emerged from the industrial city and part of its political agenda was to improve the quality of everyday life for ordinary people in the industrial city.

Work itself was transformed from handcrafted to machine-tooled, from the rule of thumb to the mechanically precise. Science was made technical rather than just experimental. The world was reimagined; electric lighting, motorized travel, and mechanical power transcended the limits of daylight, walking distance, and human strength. And this new power reframed the human and social relationship with the physical environment. The sheer power of the new industrial order broke through the previous limits to economic growth, and its seemingly limitless opportunities also upended previous ways of thought. The urban-industrial rupture created modernity.

Factory work, industrial production, and class antagonism provoked unplanned consequences in terms of health, pollution, and crime. Subsequent responses led to advances in public health, public sanitation, and social scientific understanding. The creation of public sanitation, the provision of clean water, the laying of pipes, the paying of city taxes, the emergence of urban bureaucracies, the rise of the state: all of this occurred as a result of industrial modernization and created modernity in its turn, displayed in and tempered by the city.

Consider the case of London and the famous cholera epidemic of 1854. Here, I draw upon the well-crafted book by Steven Johnson, *The Ghost Map: The Story of London's Most Terrifying Epidemic and How It Changed Science, Cities and the Modern World* (2006). The subtitle highlights the point made above. In the middle of the nineteenth century, London had a huge waste disposal problem. The city was awash in human excrement and filth. The main cause was the increase in population: it had tripled in fifty years so that by the time of the 1851 Census, it reached 2.4 million. One intellectual response posited government as the main provider of solutions. Big government grew out of the condition of Victorian cities and was epitomized in the career of such worthies as Edwin Chadwick, who was Poor Law Commissioner and member of the General Board of Health.

The new problems of the expanding cities of the nineteenth century initially eluded human ingenuity. The notion that health was a public issue led to the belief that government investment in centralized bureaucracies working with scientific experts could solve public health problems. Such responses to urban crises eventually defined the enlarged role of the state. Unfortunately, in terms of specific strategies, Chadwick got it wrong. While he rightly knew that human sewage had to be removed, his solution was to build a vast infrastructure that simply removed it from the streets and privies and pumped it into the river. This strategy was based on the mistaken belief that by removing the smells the cause of disease was also removed: the "miasma theory" of illness posited that it was the stinky air that was the cause of disease. Out of smelling range, out of danger. The medical doctor, John Snow, who prepared the map of the book's title, charted in detail the outbreaks of cholera in one particular neighborhood and linked them to a contaminated water pump. In the process the "germ theory" of disease was validated. A similar tale can be told for cities across Europe and North America. In nineteenth-century Paris, for example, Mendelsohn (2003) demonstrates how the city itself became a field and object of scientific observation. The growth of government and scientific breakthroughs as well as aesthetic sensibilities and a new form of cultural representations—significant elements of the condition of modernity— were developed and tested in the urban experience.

The shared contemporary experience of the city helped to create and refine modernity. Peter Hall (1998) highlights the connections in a series of finely detailed case studies. He sees Weimar Berlin as a general example of a cultural crucible and more particularly as one of the places that invented the twentieth century. The city was the setting for important developments in expressionism, cinema, technology, graphic art, and political theater. Cities are also innovative milieus for technological developments with far-reaching social implications. He cites Detroit 1890–1915 as the site for the mass production of automobility and San Francisco/Palo Alto 1950–1990 as the setting for the industrialization of information. He draws attention to the dream-factory movie industry of Los Angeles of 1910–1945.

New urban orders also embody and frame as well as represent and invent modernity. Paris of 1850–1870 was the place of perpetual public works, a place we will return to in a discussion of the *flâneur* in Chapter 5; London from 1825 to 1900 set out a new urban reformist path; and New York of 1880 to 1940, when it managed to handle mass movement, created the mass mobility that is the very apotheosis of the modern. Freedom, according to

Thomas Hobbes, was the absence of hindrances to motion: an understandable interpretation, given the restraints on movement in the seventeenth century when he was writing. The marvel of the modern city combines the advantages of congestion and high density with the freedom of mobility and ease of movement. A more car-dominated, individualized form of mobility was embodied in the Los Angeles of the perpetual mobility of the freeway, 1900–1980. The challenges and opportunities, obstacles and solutions of urban living in contemporary capitalism create the very stuff of the modern. I will explore the issue of modernity, mobility and the city in finer detail in Chapter 6.

New ways of thought characterized as modern also emerge from the urban context. Cities are places where ideas arise, are debated, contested, refined, and sometimes abandoned. They are places of intellectual ferment and innovation.

Stephen Toulmin (1990) extends the origins of modernity further back than the traditional account that situates it in seventeenth-century rationalism. In addition to the rationalist Baroque source for modernity, he identifies a humanist Renaissance one, grounded in the sixteenth century. This earlier agenda, exemplified in the work of Erasmus, Shakespeare, and Montaigne, is more comfortable with uncertainty, ambiguity, and difference. The later seventeenth-century agenda—and here Galileo, Descartes, and Newton are key figures—is more concerned with rationality, science, and generalities. While the first was at ease in dealing with the particular, the concrete, and the local, the latter stressed the abstract, the general, and the universal. As rationality supplanted skepticism, there was a shift in emphasis from the oral to the written, from the particular to the universal, from the local to the general, and from the timely to the timeless. The different origins are expressed in the continuing tensions in the project of modernity, especially in terms of the emphasis on the local and the particular versus the universal and the general as sources of knowledge. The period from 1920 to 1960, for example, is a modernist rupture that returned to abstract fundamentals and includes the quiet grid of Piet Mondrian's paintings, the straight lines of Gropius' buildings, and the atonal music of Arnold Schoenberg and Alban Berg. The more recent rise of postmodernism, in contrast, is a return to the earlier humanist interest. The two trajectories of humanism and rationalism embedded in the project of modernity appear and reappear, flow and counter-flow. The call of Jürgen Habermas (1992) for the unfinished project of modernity to aspire toward completion is really an appeal to the earlier humanistic tradition and its egalitarian practice of tolerance. The rise of (an explicitly termed) postmodern thought is also a reworking of the earlier humanistic emphasis on uncertainty.

The Enlightenment, the intellectual architecture of modernity, is the name given to a myriad group of trends and ideas gestating across Europe in the seventeenth century and developing in the eighteenth century. The early writers include Bacon and Descartes, Spinoza, and Locke. Later writers include Voltaire, Rousseau, Hume, Kant, and Adam Smith. Three ideas are central: the natural world can be understood through rational critical thought; the scientific method can be applied to the social world as well as to the natural world; the application of reason can lead to material and social progress.

A central feature of the Enlightenment was reason's instrumentality. "Dare to know" was Kant's motto for the movement. But even Kant worried about the effects of such clear thought on the established order. This tension between encouraging the novel in thinking without disturbing the established in social arrangements was a central one and defines Enlightenment thought across a spectrum from radical to traditional. Jonathan Israel (2001) stresses the creation of a radical Enlightenment with an emphasis on equality, democracy, secular values, and universality. He argues for the existence of a coherent European radical Enlightenment that fundamentally questioned the legitimacy of established social, political, and religious orders. A key figure in this story is Spinoza (1632–1677) who argued for an impersonal God and for the creation of secular state founded on the principle of toleration. Spinoza's goal is nothing less than for people to enjoy "continuous, supreme and unending happiness." He firmly believed in the everyday possibilities of pleasant food and drink, music, sport, and theater. Act well and rejoice, he counseled. It was the message from an enlightened urban dweller fashioning a belief system based on neighborliness, tolerance, and the civic purpose of pursuing individual happiness in a collective setting. Spinoza maps out a path that affirms the modern world in its complexity of diverse societies. Get along with others, enjoy the pleasures of the here and now in a society where individual rights are assured and happiness results from freedom. It is a philosophy of modesty, prudence, and toleration.

The Enlightenment possessed geographic dimensions. At one scale, different national concerns are sometimes asserted allowing Himmelfarb (2004) to identify British, French and American Enlightenments concerned respectively with the sociology of virtue, the ideology of reason and the politics of liberty. The precise categories are debatable but Enlightenment thinking did develop in specific places. Cities were especially fertile places for the generation and refinement of new systems of thought about the connections between individual rights and social obligations, between self-interest and the common good. Ideas circulated in the newspapers and pamphlets, were discussed in clubs and coffee houses and were tested and contested in debates and discussions. Nowhere and no time is this more apparent than in the urban bias of the Scottish Enlightenment.

All urban creativities seem to be like bolts of lightning; they come from a dark sky, are immensely powerful yet tantalizingly brief, as with Athens in the time of Pericles, Quattrocento Florence, or fin de siècle Vienna. Edinburgh moved quickly from a provincial backwater to a leading European intellectual powerhouse of a city. In 1696 a nineteen-year-old theology student, Thomas Aikenhead, was found guilty of blasphemy. His crime: he said that the Bible was not the literal word of God but a set of romances. For this, in Presbyterian Edinburgh at the end of the seventeenth century, he was hanged. In little more than sixty years after this, the city was hosting Benjamin Franklin, and Voltaire was praising the taste of the Scots. From theocracy to the rule of reason in less than sixty years. The city is one of the premier centers for the invention of modernity (Herman 2001; Buchan 2003).

From a large cast of characters, a few individuals stand out. Francis Hutcheson (1694–1746) and Lord Kames (1696–1782) codified the beginnings of social studies. Mankind was brought under the scope of examination. Hutcheson countered the Hobbesian view of life as short, nasty, and brutish with a society based on maximizing individual rights and the pursuit of happiness. Similar ideas were to echo in the American Revolution. While Hutcheson posited an innate moral sense, Lord Kames suggested a more hardheaded view. Protection of property was a primary source of motivation. He espouses what later Marx transformed into historical materialism; the idea that human societies evolve through stages of economic activity; from hunting–gathering through herding and agriculture to urban societies. In each stage new rights and obligations developed. Kames realized that there was an economic base to social arrangements and relations.

David Hume (1711–1776), "modernity's first great philosopher" (Herman 2001, p. 169) carried on the line of thinking opened up by Spinoza. There was no God, no divine plan; self-interest was the motivating force. Society was a way to channel passion and self-interest. Self-interest was also at the heart of Adam Smith's work, the area we now call economics. Smith (1723–1790), a close friend of Hume, combined the self-interest of individuals with the division of labor in modern economies into a model of a dynamic society. Free individuals acting out their self-interest could provide collective benefits. Adam Ferguson (1723–1816) was not so sanguine about the cultural effects of capitalism; he saw a civil society, a term he invented, that was based on individualism and avarice as leading to tyranny.

The ideas shaped and tempered in Edinburgh spread. Ferguson's ideas permeated the German Enlightenment and especially influenced the work of Herder, Schiller, and Hegel. Ideas were taken across to America and exported elsewhere abroad as British imperialism, commercial and intellectual, triumphed. And even in his wildest dreams, Adam Smith could not foretell how his ideas on the role of the market would come to dominate not only economic analysis but also political thinking and social beliefs around the globe.

There were also developments in medicine, technology and science. Joseph Black (1728–1799) discovered latent heat and carbon dioxide. He was a friend of James Watt (1736–1819) an instrument maker at the University of Glasgow who improved the efficiency of the steam engine. His name lives on in the measurement of the rate of energy conversion: one watt is equal to one joule per second. James Hutton (1726–1797) was the first geologist to suggest that the earth was older than the standard Christian view of 6,000 years old. He suggested it was, based on the geological evidence, millions or even billions of years old. The finding undermined centuries of Christian doctrine and, like the theories of Copernicus and Darwin, dislodged humans from the center and beginning of creation. The concept of deep time displaced the shallow human-centered time of Christian doctrine. In less than a hundred years, the city moved on from hanging blasphemers to lauding the scientists who overturned centuries of received Christian doctrine.

There was vigorous debate among all the participants. They influenced each other directly, interacting socially and professionally in clubs and societies, social events, dinners, drinking bouts, and literary salons. The sheer concentration of genius is evident in the fact that all of the eight people mentioned above were alive at the very same time, the period from 1736 to 1746. For slightly more than quarter of a century from 1750 to 1776, all but one, Francis Hutcheson, were alive. They aided and helped, disagreed and fought. Black, Hutton, and Watt were friends. Hume and Smith were best friends. Black was Hume's doctor and regularly corresponded with Smith about Hume's health. Hutton's ideas on deep time were first presented at the Royal Society of Edinburgh in 1785. Adam Smith was in attendance.

Their collegiality also hosted vigorous disagreement. Hume for example saw Ferguson's writings as romantic primitivism that lauded the past rather then celebrated the modern present. Individual works developed out of dialogues, and these dialogues were heightened and concentrated in the tight urban space of Edinburgh. In his autobiography published in 1777 David Hume wrote, "I removed from the Country to the Town, the true Scene for a man of letters."

The Scottish Enlightenment centered in Edinburgh, also encompassing Glasgow, grew out of a particular urban experience but also gave shape to a more general urban experience. The general problem shared by all urban dwellers in the modern age was how could they live together without the benefit of an eternal deity who rewards or punishes our behavior and shapes our destiny? If we undercut traditional beliefs that we are acting out God's plan or operating under divine principles, then we have to look at people in the clear light of day. Looking at them with the rational gaze of everyday urban encounters, they are no longer pawns in divine drama but active agents of their society, makers of their own predicament. Looking at them closely in the city streets reveals passion-filled, self-interested individuals. How can these atomistic elements be combined? How can self-interest can be harnessed for the collective good? Individual rights and self-interest, the Enlightenment philosophers argued, create free societies. But we also have obligations. Authority enhances individual freedoms through the enforcement of our obligations. It is a supremely confident and optimistic view. The cold-ness of a Godless world is replaced by the warm vitality of a human society of free individuals whose pursuit of self-interest leads to collective improve-ment. It is the view of the world from the perspective of the enlightened, self-interested individuals brought together in commercial relations, social contracts, and urban living. People are social animals. For the Enlightenment thinkers of Edinburgh, social solutions could not lie in received dogma, living out outmoded moral codes or maintaining traditional hierarchical societies. The answer, according to those who embraced modernity, lay in freedom and tolerance, skepticism and rationalism, a belief that observation, description, comparison, and classification reveal how things really are. It is a philosophy for the urban here and now, shaped in the cosmopolitan context of city life. When the French philosopher Duclos wrote in 1750 about Paris that "those who live a hundred miles from the capital are a century away from it in their modes of thinking and acting," he reminds us of the urban bias of modernity (quoted in Gay 1969, p. 4). The modern was an urban experience and modernity was invented in the city.

Multiple modernities

There never was just one modernity, and there were always multiple interpretations. There were dual origins in Renaissance humanism and there were the radical and more moderate forms of the Enlightenment, exemplified in the contrast between Spinoza and Leibniz (Stewart 2006).

Modernity was not all just life enhancing improvements in medicine and sophisticated philosophical discussions: it was also about the imposition of order and the exercise of power. There was a dark side to the Enlightenment and modernity as there was to the urban experience. Reason and enlightenment also had its counterparts in the definitions, disciplining, and control of unreason and the unenlightened (Foucault 1967). There were spaces of exclusion and sites of control as well as clearings of rational discussion. The two were linked. Chris Philo (1999) tells the intriguing tale of Robert Fergusson, a poet of the Scottish Enlightenment. After an onset of mental illness he was committed to a series of institutions, ending his days in October 1774 in a lonely cell in Edinburgh City Bedlam, an institution for the mentally ill indigent. The cultivation of reason in Edinburgh also had a corollary in the confinement of those considered unreasonable. The places of reason were connected to the places of unreason.

Order and reason had their undersides in the political construction of order and the disciplining of the unreasonable. For the social theorists such as Horkheimer and Adorno, writing in the aftermath of World War II, there was a dark heart to the Enlightenment. Overdetermined rationality allied to political forces no longer in service to the ideal of a liberal state constitutes the failure of Enlightenment, leading inexorably, according to some commentators, to the guillotine and the gas chamber.

For some, the bankruptcy of modernity heralds a new postmodern condition that lacks a belief in progress, reason, and universal truth. There is growing resistance to what were traditional patterns of Eurocentric dominance, of patriarchy and technical expertise, resistances to the notion of modernization and a distrust of metanarratives of science. Others deny the validity of the term postmodern, arguing that we are still grappling with modernity, although we may be in a later more reflexive stage. Some sociologists, most notably Ulrich Beck, Anthony Giddens, and Scott Lash (1994), propose a second wave of modernity, a flexible modernization that is a responding to the first wave that brought democracy, civil rights and the welfare state. Increasing individualization, decline of the state in the face of globalizing forces and a rising concern with managing risk and ensuring sustainability marks this new wave. This second modernity underscores the inadequacy of existing institutions that grew out of the first modernity. Zygmunt Bauman coins the term "liquid modernity" to refer to new forms of nomadism and uncertainty. Anthony King uses the term "multiple modernisms" with reference to the inflected practices of modernism across the world. He identifies colonial modernity as well as an indigenous modernity, and the hybrid modernity that results from the collision of the two. King (2004, 2009) goes on to ask the question: What are the signs of a modern city? He identifies five that combine social and spatial modernity: a self-consciously "world" city, a fascination with tall buildings, the work of signature architects and spectacular architecture, multiculturalism, and modern ways of life. The pursuit of these signs is key to understanding urban change. Not all cities pursue all these signs of modernity with the same zeal. There is the case of Dubai, which has invested most heavily in creating modernity through architecture spectaculars. Consider the pursuit of the tallest building. Plotting its urban geography tells much about the changing center of gravity of metropolitan modernity. In 1931 the Empire State Building in NYC, topping out at 381 meters, was the tallest building in the world. Then the designation shifted to Chicago (Hancock Building), back to New York (World Trade Center) and in 1974 back to Chicago with the constructing of the Sears Tower, now the Willis Tower as the tallest in the world. The competition increases as developing cities in the Middle East and East Asia seek the designation: Kuala Lumpur, Taipei, and Shanghai. In 2010 the Burj Khalifa building in Dubai, its 160 floors rising to 828 meters, was the world's tallest building. Built at an estimated cost of $800 billion, the project was just the culmination of a series of architectural spectacles in the oil-rich emirate. There was the mixed residential and entertainment complex of Palm Island built on man-made islands in the shape of a palm leaf; World Islands, a luxury residential complex on artificial islands shaped as a map of the world; and the Mall of the Emirates, which included under its roof in the hot, dry desert air an indoor ski slope. The conspicuous display of excess was not matched by a commitment to a broad multiculturalism; regulations welcomed the foreign rich and flight capital from around the world but not the poor migrant workers who built the gleaming towers, often in difficult conditions with limited rights. Modernity was embraced but not all the elements of modernity were embraced equally. Dubai's growth and excess, fueled by seemingly endless oil revenues, crashed in 2009 when the sovereign wealth fund, Dubai World, which had bankrolled the many projects, was discovered to be in deep debt.

I will characterize the term "metropolitan modernities" to try to capture some of the complexities. These modernities are fashioned in complex spatial contexts. I will discuss three: a triad of metageographies, space–time warpings, and urban performance space.

Three metageographies

Metropolitan modernities are fashioned in a complicated space resulting from the foldings, intersections and imbrications of three metageographies. The first is shown in Figure 3.1, a standard world map that depicts the world's land surface partitioned into nation-states. In many such maps color-coding often highlights individual countries; the world imagined as a series of separate homogeneous countries. I will refer to this as the state-centered metageography. Despite waves of globalization and reglobalization, the nation-state is a sturdy survivor. As I argued in Chapter 2, the state is not so much replaced but restructured in the Late Modern Wave of Globalization and indeed some sections of the state apparatus are strengthened. The nation-state survives as a source of identity, a forum of political debate and an important economic agent. Nation-states are units of monopoly political control. While there are still disputed borders, competing land claims, and unresolved issues of legitimate sovereignty, nation-states cover the surface of the globe. With some notable exceptions such as Antarctica and deep-sea regions, the world's surface is a jigsaw of separate but connected national surfaces. The global metageography of political surfaces and boundaries is nation-states.

The state-centered perspectives dominate much of our understanding of the world. The Gross National Product of national economies, for example, is a standard economic measurement. Nation-states are basic building blocks in popular understandings of the world.

Immanuel Wallerstein proposes an alternative metageography to this state-centered model in an influential three-volume study that describes a world system comprising core, periphery, and semi-periphery (Wallerstein 1974, 1980, 1989). Wallerstein's focus is on the establishment and maintenance of global capitalist order that revolves around the dominant core economies of Western Europe and later North America. Economic transactions are marked by an unequal exchange of cheap raw materials from the periphery to the core and the export of manufactured goods from the core to the periphery. This essential dynamic explains the colonial expansion of core countries as they sought to establish monopoly control over regions that were then transformed into colonial peripheries to feed and sustain the core economies by supplying cheap raw materials and purchasing manufactured goods. To take just one example: in the nineteenth century, Britain was a core country that incorporated much of India into commercial domination and political control. Raw cotton grown in India was shipped to Britain and manufactured in Lancashire cotton mills and then exported abroad. The indigenous cotton manufacturing industry in India was destroyed while the domestic cotton industry in Britain

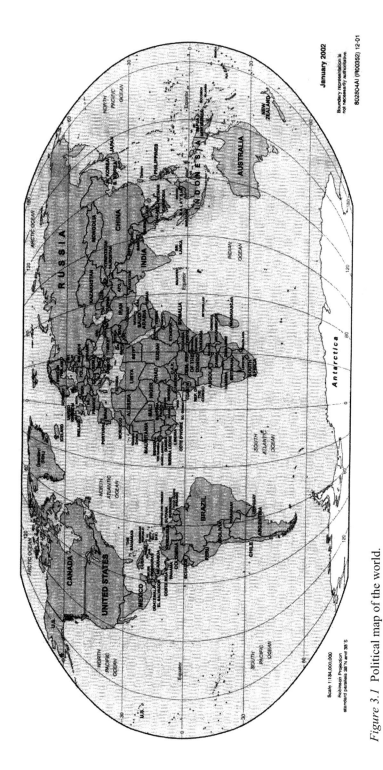

Figure 3.1 Political map of the world.

Source: 2010 http://commons.wikimedia.org/wiki/File:CIA_Political_World_Map_2002.jpg (accessed on December 10, 2010).

was a leading force in the Industrial Revolution that propelled national economic growth and an imperial project. This core–periphery structure allows us, indeed forces us, to globalize such issues as the Industrial Revolution, presenting them less as unique national experiences and more as part of a global drama. The model is dynamic. The category of semi-periphery is something of a transitional condition situated somewhere between the dominance of the core and the relative weakness of the periphery. Some countries can move from periphery toward the core. In recent years the rise of Brazil, China and South Korea is part of a global economic restructuring in which these countries, or to be more precise, specific metropolitan regions of these countries, export manufactured goods rather than raw materials. And even some strategic raw primary producers such as the oil-producing countries of the OPEC can use their collective interests and combined power to increase prices and force redistribution in global wealth. The world system model of core–periphery has the enormous advantage of globalizing historical events and adding a much-needed political economy to standard historical narratives. Modernity is thus firmly situated in a wider historical and deeper materialist discourse.

The core–periphery model is capable, with some tweaking, of also being deployed at different scales. At the national scale it is useful as a way to understand marked regional inequalities. Take the case of China, less the homogeneous surface that the political world maps imply, and more a coastal core of rapidly industrializing cities with a huge rural peripheral interior. The vast rural to urban migrations are similar to the international migration paths followed by people from poor peripheral countries who move to rich core countries. The terms core and periphery can apply to an individual country as well as to groupings of countries. It can also be used at the scale of the individual city. The model allows us to imagine city regions structured around cores of globalized industries and connected communities. The peripheries are the informal economies, the squatter settlements, and the peripheral shantytowns. The core–periphery metageography (and mesogeography of individual states and microeography of city regions) deals with spatial relations of power. If the nation-state perspective tells us about spatial differences, the core–periphery highlights spatial power relations.

In Figure 3.1 each country is pictured as one surface. Large and small, the countries are envisioned as a homogeneous unit. All of the US is one country as is Australia, China and Russia. However, in the wake of recent changes and especially of the Third Urban Revolution, certain regions in different countries are often more alike than other regions in the same country. Across the surface of the earth global cities in different countries are as similar and as linked, sometimes more so, than rural areas in their own national territory. The state-centered world picture tends to downplay or ignore this import-ant feature of our contemporary world. Another metageography is revealed in the night-time image of lit areas across the globe produced by NASA at http://science.nasa.gov/media/medialibrary/2000/11/15/ast15nov_1_resources/earth_lights.jpg). This image, based on satellite images of the earth at night, is not a single satellite image but a composite. The image highlights, quite literally, another global spatial order, the metageography of large, well-lit city regions. In comparison with Figure 3.1, it reveals points rather than surfaces. The northeast US, much of Western Europe, and most of Japan is brightly lit. Look at Australia. On a political world map the entire country would be represented as a single surface. In the NASA image the dark interior is contrasted with the coastal cities; on the east coast and moving clockwise are Brisbane, Sydney, Melbourne, and Adelaide while on the south-west coast Perth sits alone on the edge of the dark Pacific Ocean. Away from these urban pools of light there is comparative darkness. Earlier in the chapter I commented on the urban bias of the Enlightenment; here is its contemporary equivalent, the city as pinpricks of light against a dark background.

These urban points are connected by flows of goods, people, capital, and ideas. Some of the flows have been identified and measured. A substantial body of material has emerged from the work of Peter Taylor and his colleagues at the Globalization and World Cities (GaWC) research network. The website (www.lboro.ac.uk/gawc/) lists data sets as well as more than 350 research papers and is an indispensable guide to the metageography of urban networks. In 2000 they collected data on the distribution of 100 global advanced producers service firms, including accountancy, advertising, banking/finance, insurance, law, and management consultancy, across 315 cities. They analyzed the resultant data matrix to identify a global urban hierarchy. In 2008 they extended the analysis to 175 firms in 525 cities. The result was a fivefold hierarchy that identified cities as alpha, beta, gamma, high sufficiency, and sufficiency. Figure 3.2 is a cartogram of alpha cities in the network of advanced producer services. Note how New York (NY) and London (LON) dominate. NYLON is an important pivot in the global networks, and its dominance reflects the historical legacy of London as the center of the British Empire and the continuing importance of New York as the financial center of America's more informal empire. The map also hints at the rapid rise of new cities into the top tier as national economies make their way from periphery to core. In 2000 Shanghai was three steps below in the alpha minus category while Beijing was only a beta city. In 2008 both Shanghai and Beijing were classified as alpha plus, only one step below NYLON. Both cities are moving into the top tier as China's economic growth, both absolute and relative to the rest of the world, continues apace. The diagram highlights the centers of new metropolitan modernity, the rapidly growing cities of the Far East.

Lisa Benton-Short and colleagues also sought to identify a global network, but their work was based on flows of people. They looked at immigration into cities around the world and established an index based on the percentage of foreign-born, the total number of foreign-born, the percentage of foreign-born not from a neighboring country, and the diversity of immigrant home countries (Benton-Short et al. 2005). The result was a threefold division into alpha, beta, and gamma cities.

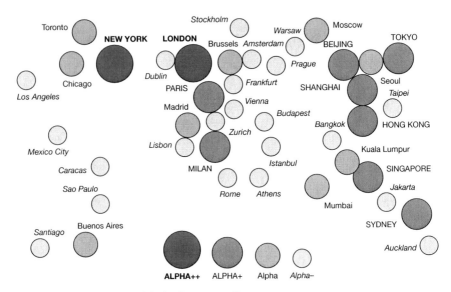

Figure 3.2 The GaWC global urban network.
Source: www.lboro.ac.uk/gawc/world2008c.html (accessed on 15 December 2010).

There are similarities and differences between the GaWC and the Benton-Short et al. results. Confining comments to the top level and looking first at similarities: some cities appear in the same category in both analyses. New York and London are alpha cities and sit atop the apex of both hierarchies. Other shared alpha cities include Toronto, Los Angeles, Sydney, and Amsterdam. Differences: while Miami, Melbourne, Vancouver, and Dubai are considered pivotal points of global migration reaching alpha status, in the advanced producer service category they only make it to beta status.

Some of the differences relate to issues of data and data availability. But they also refer to the nature of the global urban network. Networks vary according to the flow. Some flows "pool" in some cities rather than others. Advanced producer services, for example, which can be considered command functions of the globalizing economy, are still concentrated in just a few cities because of the need for social interaction in global financial business deals. Trust, contact networks and social relations play pivotal roles in the smooth functioning of global business. Spatial propinquity allows these relations to be easily maintained, lubricated, and sustained. Select cities are the sites of dense networks of interpersonal contact and centers of the important business/social capital vital to the successful operation of international finance. Global service corporations are adept at producing their own commodities, including new financial products, new advertising packages, and new forms of multi-jurisdictional law. All of these depend upon specialized knowledge. The coalescence of a range of expertise produces state-of-the-art commodities to meet the specific needs of clients. In order to be able to put together such packages, firms need to be in knowledge-rich environments. Face-to-face contacts between experts are facilitated by the clustering of knowledge-rich individuals in cities such as New York and London. Cities at the apex of this particular network are "privileged sites" housing the "knowledge elite" that enact the economic reflexivity crucial for economic success. Reflexivity and networking are at the heart of understanding global cities as places where people, institutions and epistemic communities work to establish and maintain contacts. More importantly, these communities act as crucial mediators and translators of the flows of knowledge, capital, people, and goods that circulate in the world. A GaWC alpha global city attends to the heterogeneous global space of flows, lending intelligibility and translatability to otherwise incommensurable materials, such as credit rating that mediates diverse banking systems and global law that translates between different jurisdictions.

The flow of people, while matching the connections in command functions, also has slight differences. National regulations concerning immigration, the demand for labor, and the relative openness of societies to foreign migrants all play a part. The sheer need for labor, whether the global talent pool of specialized knowledge experts or cheap unskilled labor varies throughout the urban network. The result is a similar but not identical match of command centers with immigration hubs.

There is no one fixed global urban hierarchy but more of a space of flows, to use a term coined by Manuel Castells (1996); a global urban network with different configurations of hierarchy depending on the flows. Money, people, ideas, goods, and practices do not just flow through the urban network, they are transformed in the process. To be more accurate we should use the term "space of transformative flows." Consider flows of people. The movement through the urban network is not a simple geographical movement; it involves cultural exchanges. These can refer to the new work habits and job practices of the temporary worker as well as the complex cultural transformations of long-term migrants as they adapt to new milieus and in turn transform their surroundings. Rather than mere transfers, flows along the urban networks are transformative experiences, even in the flows of inanimate things such as money, goods and services. A small amount of money in London or New York when remitted back to Ghana or El Salvador can become the source for land and house purchase, enabling a new business or paying for better schooling. As I remarked in Chapter 2, cultural globalization is not a process of homogenization but a richer more complex form of hybridization, mimicry, and continual transformation. Now we can see that in large measure cultural globalization is the movement of ideas and practices through the global urban network, transforming the flows and the nodes in a continual incessant process of transfer and change. Flows through the urban network change the medium and the networks; as people adapt, ideas are tweaked, money is reimagined, and commodities are reappropriated.

While the aggregate analyses are immensely useful, they also have their biases. The cities of the West tend to be privileged sites for investigation. Intensive fieldwork in less privileged cities reveals a different story as in the detailed case study of Accra by Richard Grant (2009). As the capital of the West African nation of Ghana, Accra does not figure highly in the standard measures of globalization. On the GaWC classification, it is listed on the fourth level of high sufficiency, well down the list. At first blush then, just another city in remote Africa, largely cut off from global flows. Grant presents a very different picture. He highlights three key moments of globalization: the selection of the city as colonial capital of the Gold Coast in 1877, when it became a gateway for goods from the British Empire and a transmission point, taking out the wealth of interior Africa; the end of colonial rule in 1957, creating a more inward focused economy as industries were nationalized, trade tariffs were erected and foreign investment was heavily regulated; and a massive economic restructuring inaugurated in 1983 under a World Bank/IMF structural adjustment program that ushered in a neoliberal agenda. Foreign companies now play an important role in the urban economy, constituting one-third of all firms. The urban form itself was transformed. A new foreign-orientated central business district was established close to the airport. There was also the rise of gated communities, especially since 1991, housing mainly transnationals and expatriates. The city's economy is fueled by remittances from Ghanaians abroad. Returning home, these repatriates often play key roles in transnational operations and flows. But even those stuck in the city also have global connections. Grant shows that shantytown dwellers are also connected to transnational NGOs working with them to frame their demands and need. This detailed case study reveals the pervasive and multiple connections with the global space of flows of people, money, discourses, practices, and ideas. Accra is revealed as a networked city with globalization being imposed as well as embraced, transforming from above as well as changing from below. No African cities make the top tier of GaWC cities and yet, as the Grant study shows, Accra is a globalizing city firmly connected into global flows.

The three metageographies are not independent of each other. Among the many points of overlap, let me mention just a few. In terms of the connection between the urban network and the core–periphery metageographies there is a historical legacy of colonial cities. Urban centers were transmission points in the construction and maintenance of core–periphery relations. Cities were gateways for trade and sites of domination and control. Throughout the world existing cities were turned into colonial cities of control and extraction points in core–periphery trade relations. Throughout much of Latin America, for example, the Spanish empire ruled in and through cities that subsequently dominated postcolonial societies and economies.

Imperialism operated through cities. Colonization involved a rewriting of urban space to show who was in control and who was controlled. Timothy Mitchell (1988) identifies three broad strategies for the framing of colonial states: producing a plan of urban segmentation that includes racial segregation; creating a fixed distinction between inside and outside; and constructing central spaces of observation to keep an eye on things and to show the presence of colonial power. In Accra, as in many other cities, new areas of European settlement were created sharply demarcated from the "native" areas. Laws barred "natives" from living in areas reserved for whites. Further north, Tunis came under French rule in 1881. A new town was laid out on a gridiron plan right beside the old Islamic city. The garden suburbs of the French colonizers were in rich contrast to the spontaneous settlement at the edge of the city that housed the huge influx of migrants from the countryside. The planned differences in the colonial city between public and private and between home and work stood in stark contrast to the cellular structure of the old Islamic city. The French built neoclassical buildings and created straight boulevards and symmetrical squares as a direct response to the tight and convoluted street pattern of the old Arab town; Gallic rationalism counterpoised to Islamic chaos. When the French established Hanoi as the capital of French Indochina in 1887, they created a city in their own image. Long, wide boulevards, reminiscent of Haussmann's Paris, were cut through the city, public parks were laid out, and new buildings were constructed that would not look out of place in Paris or Bordeaux.

There have been two subsequent processes. On the one hand, there is a post-colonial appropriation that involves renaming and reoccupations, inscriptions and erasures. In Accra the black elite moved into areas formerly reserved for Europeans. I will consider the dynamics of a postcolonial city in more detail in some of these processes in Chapter 9. On the other hand, there is the continuing legacy of the dominance of many former colonial command centers in the political life and economic space of the independent nations. Accra, Tunis, and Hanoi still dominate their respective countries.

The core–periphery relation structure also affected the imperial core cities, whether it be in the archives of imperial/colonial domination such as botanic gardens that house "exotic species," museums that contain treasures from "around the world" and "international" universities that narrate the world, or in the form of colonial legacies such as the North African population in French cities, Puerto Ricans in New York City, or Jamaicans in London. As more second and third generation immigrants from the colonized periphery live and work in the cities of the colonizing center, new forms of identity, resistances, and hybridity emerge. Issues of citizenship, national identity, and urban community are as important in the cities of the core as they are in the periphery.

The system of nation-states and the global urban network are not simple homologies. We can consider some of the differences by comparing globalization indices for cities and countries. Every few years the journal *Foreign Affairs* lists countries on a globalization index that measures political engagement, personal contact, technological connectivity, and economic integration. In 2007 the top twenty countries were Singapore, Hong Kong, the Netherlands,

Table 3.1 National and city globalization indices

Foreign affairs country index	Main city	GaWC rank
1 Singapore	Singapore	5
2 Hong Kong	Hong Kong	3
3 Netherlands	Amsterdam	23
4 Switzerland	Zurich	22
5 Ireland	Dublin	26
6 Denmark	Copenhagen	58
7 US	New York	2
8 Canada	Toronto	14
9 Jordan	Amman	109
10 Estonia	Tallinn	121
11 Sweden	Stockholm	33
12 UK	London	1
13 Australia	Sydney	7
14 Austria	Vienna	35
15 Belgium	Brussels	15
16 New Zealand	Auckland	40
17 Norway	Oslo	54
18 Finland	Helsinki	56
19 Czech Republic	Prague	34
20 Slovenia	Ljubljana	101

Switzerland, Ireland, Denmark, the US, Canada, Jordan, Estonia, Sweden, the UK, Australia, Austria, Belgium, New Zealand, Norway, Finland, Czech Republic, Slovenia. The index is biased toward the smaller richer countries as well as the very rich big countries such as Australia, Canada, and the US. Table 3.1 ranks the country with the equivalent GaWC rank of the most globalized city within the national territory. Thus, while the UK ranks twelfth in terms of the *Foreign Affairs* country index, the GaWC index puts London at number one. Given the differences in the data and methodologies the comparisons need to be treated with some care. But even with this proviso it is clear that we can identify the *globalized country/globalized city*, particularly marked in the city states of Singapore and Hong Kong where country and city are collapsed into one global hub. There is also the *globalized country/less globalized city* including the small, often rich, countries where the city rank drops below the country rank—Denmark/Copenhagen, Austria/Vienna, Slovenia/Ljubljana, Estonia/Tallinn, and Finland/Helsinki. Here are small rich countries with relatively high levels of wealth across the country that allow national integration into global networks, but the relatively small size and resultant lesser global connectivity of the city drops the urban ranking. Then there are the silences from this table, indicating *globalized city/less globalized country*. The most obvious example is China, which only registers sixty-sixth in the *Foreign Affairs* globalization index, but Shanghai and Beijing rank nine and ten respectively in the GaWC rankings. These cities are global hubs in the network of advanced producer services in a national sea of rural poverty.

In terms of the connections between the urban network and nation-states; we can identify a number of issues that can be summarized under the headings of *big city/small country* and *small city/big country*. In the extreme case of *big city/small country* a single city can dominate an entire country. This tends to occur in small and medium sized countries especially those with recent histories of imperialism or colonialism. In the case of London, Paris, and Vienna for example, the size of the cities refers to previously larger and more extensive imperial/colonial territories. In the early twentieth century, London was the capital of an empire that stretched around the world. In 1922 it was the nerve center of an empire with a population of 458 million. Ninety years later it was the capital of a country with a population of only sixty-five million. While the empire shrank, the city continued to grow. I will explore these issues in greater detail in Chapter 8.

In the case of *small city/big country*, cities are not in a dominant national position. This can occur in a relative sense when there is competition between cities. Take the case of Australia. We last left it in the darkness with only the coastal cities throwing enough light into the air to be recorded by satellites. While the individual cities are large, the populations of Sydney and Melbourne, for example, are 4.5 million and 3.9 million respectively out of a national population of 22.3 million, no one city dominates the national scene as in the case of London and the UK. Here the struggle for global recognition and global competitiveness between them is keen and constant. The imagining of the city is always in position to the national urban Other as well as to other globalizing cities around the globe.

Small city/big country can also occur in the absolute sense when cities are small in relation to the country. In the Third Urban Revolution this is a rare exception. It tends to occur then in unusual political circumstances that override the usually dominant economic forces that lead to increased urban concentrations. As examples I will cite the case of two totalitarian regimes that actively sought to undermine their largest city. Pol Pot, the murderous head of the Khmer Rouge came to power in Cambodia in 1975. In April of that year he swept into the capital city of Phnom Penh, the historical capital of the country since the fifteenth century. It was by far the largest city in the country and the site of French colonial authority from 1870 until a shaky independence in 1953. The large sophisticated city with a diverse population of over two million was an affront to the Khmer Rouge who preached a form of rural socialism that quickly turned into a savage killing machine. The regime wanted to undermine the city's importance and annihilate the city dwellers. The regime pursued a radical program of relocating people from the city, abolishing money and private property. The Khmer regime waged a war against modernity. One specific aim was to get rid of the "new" people of the cosmopolitan city; those with glasses were especially picked upon as a sign of their education, and turned into "old" people through forced agricultural labor. Many citizens were tortured and killed. Almost two million died through exhaustion and starvation or were killed during the reign of Pol Pot. It was only in 2010 that the city regained its pre-Khmer population of two million.

Big cities are especially repellent to austere authoritarian regimes that dislike their size, their diversity and the opportunity they create for the expression of dissent and the mobilization of resistance. The military junta that controls Burma (Myanmar) continually tries to undercut Rangoon's national prominence and especially its role as a possible stage for protests against the regime. The city has a population of 4.3 million. In 2006 they relocated the capital to Naypyidaw, a greenfield site almost 200 miles north of Rangoon. Construction began in 2002 in an act of government policy only possible in a repressive regime. As the journalist Siddarth Varadarajan (2007) noted after a visit:

> Vast and empty. Burma's new capital will not fall to an urban upheaval easily. It has no city center, no confined public space where even a crowd of several thousand people could make a visual, let alone political impression . . . the ultimate insurance against regime change.

There are thankfully few examples of Pol Pot's inhuman savagery or the Burmese military junta's thuggery. However, in periods of rapid change and economic dislocation and political uncertainty, the countryside can be mobilized in political discourses in which the city, as the site of the Other, becomes the source of all that is bad and wrong; I have discussed this bias against the urban in some detail (Short 1991).

There are many folds and overlaps in among the three metageographies. In the case of *big city/small world* we see the very large global cities that sit atop the global urban networks. These are centers for global command functions, the hubs of international migrations, and centers of representations, sites of cosmopolitanism. The economic global linkages and cosmopolitanism are central to the character and regular representations as they boast and boost their international competitiveness and tolerant cosmopolitanism, not simply the result of more foreign-born but also in terms of an active conversation with the foreign Others. These are the principal sites of what is described as transurbanism, a global urban citizenry whose sense of identity is based both on global connections and the production of locality, a cosmopolitanism that is sensitive to and participates in local traditions (Appadurai 2002).

Outside the top tiers of the urban hierarchy we are in the realms of *small city/big world*. The emphasis on identifying "world" and "global" cities tends to concentrate attention at the apex of the global urban networks and ignores how global flows act in and through all cities. There is a real need for extending the globalization/city research nexus beyond the usual suspects of very big rich cities. I have extended this argument elsewhere (Short 2004a) to make the case that the term globalizing city is a more accurate term for all the cities in the world, linked as they in a complex web of flows, connections and transactions. I will develop this argument in greater detail in Chapter 9 when I discuss the small city of Alice Springs/Mparntwe and its role in the active construction of a postcolonial modernity.

There are also cases where the relations between the city and the global are frayed and broken. I have called these the black holes and loose connections of the global urban network. To identify very large cities only loosely connected to the rest of the world I combined two data sets from the early 2000s. The study is more fully documented in Short (2004b). The first is the population figures for major agglomerations in 2002 available in Brinkhoff (2010) who provides the most accessible and up-to-date population figures for urban agglomerations around the world, ranked by population size. The population data for individual cities were compared to the GaWC 2000 data. Eleven cities were identified that met three criteria: they each had a population of over three million, were not identified by GaWC as a world city and did not share their national territory with a world city. They are listed in Table 3.2. They ranged from Tehran with a current population of 12.8 million to Chittagong with a current population of 4.6 million. There are a number of reasons behind these very large cities' non-world city status. I will posit five: poverty, collapse, risk aversion, exclusion and resistance.

Some cities, despite their size, are so poor that they do not represent a market for advanced producer services. They are the black holes of advanced global capitalism with many people but not enough affluent consumers or complex industries to support sophisticated producer services. Approximately eight of the eleven cities in Table 3.2 are in what are classified as low-income countries by the World Bank, and three (Tehran, Baghdad, and Algiers) are in the low–medium category. These cities are in some of the poorest countries in the world. Kinshasa is the capital of a Congo whose gross national income per capita in 2009 was a pitiful $300. Dhaka and Chittagong are in Bangladesh,

Table 3.2 Black holes

City	Country	Population (million)*	GNI per capita (US$)*	Risk rating
Dhaka	Bangladesh	13.6	1,600	Significant
Tehran	Iran	12.8	12,900	High
Kinshasa	Congo	8.9	300	Extreme
Lahore	Pakistan	8.5	2,600	High
Baghdad	Iraq	6.6	3,600	Extreme
Khartoum	Sudan	4.9	2,300	Very high
Rangoon	Myanmar	4.7	1,100	High
Chittagong	Bangladesh	4.6	1,600	Significant
Abidjan	Ivory Coast	4.4	1,700	Significant
Algiers	Algeria	3.1	7,000	High
Pyongyang	North Korea	2.6	1,900	Very high

* 2009 data

a country where the gross national income per capita was $1,600. The comparable figures for the US and UK are $46,400 and $35,200, respectively. Many of the cities' populations are poor, living on the margins. These cities lack a significant middle class and an advanced urban economy. Not requiring the services of global producer service firms they are excluded from GaWC world city status. It is not legitimate to write of urbanization without global-ization, since all cities partake in some form of global connections. Urbaniza-tion with only selective economic globalization is perhaps a more accurate term. The dominant form of their globalization is usually in the form of migration flows and remittances.

There are cases of not only endemic poverty but also of catastrophic decline where there has been an almost complete collapse of civil society. In recent years, Khartoum and Kinshasa, for example, have witnessed the decline of the rule of law and social anarchy. War and social unrest have been the norm rather than the exception. These two cities represent cities that have internally collapsed and are abandoned or bypassed by global capitalism. Sustained social disruption reinforces the global disconnect.

Poverty and social anarchy do not explain all the cases. Some cities are bypassed because of risk aversion by capital investors. There are many global metrics of risk. One private company produces a risk rating of national econ-omies (World Markets Country Analysis 2002). For 2002, they ranked 185 countries from insignificant risk (Luxembourg rated 1) to extreme risk (Afghanistan rated 185). The risk categories are also shown in Table 3.2. All the cities were rated as significant, high, very high or extreme risks and their numerical values were skewed toward the extreme risk end of the index. The perception of risk can be self-fulfilling. High risk puts corporations off from establishing connections, which in turn increases the high risk rating. National ideologies also play a role. Khartoum, Pyongyang, Rangoon, and Tehran, for example, are all cities in countries where national ideologies have not encour-aged global economic connections to the advanced capitalist economies. In both Khartoum and Pyongyang the population fell from 2002 to 2009, while that of Rangoon remained static, rare events for large city regions. Tehran stands out for being in a country significantly wealthier than the other black holes, the result of the rich oil resources; here ideology trumps the resource bounty. The black holes are made up of the poor city, the collapsed city, and the excluded city.

Space–time warpings

A crucial ingredient of modernity is the experience and representations of time and space. In the long wake of James Hutton's initial work, we have a scientific understanding of deep time as well as the experience of more shallow time. The understanding is not universal. There is still a sturdy belief in the short time of creationism. According to a 2007 Gallup poll, 48 percent of adults in the US do not believe in evolution. Almost 70 percent of people who described themselves as Republicans do not believe in evolution. Deep time, the opinions of creationists aside, is embedded in the evolution of the earth, the geologic time of millions of years and the long time of human evolution. Shallow time can be broken down into shorter-term cycles from the life cycle of the average human to the daily rhythms of our everyday lives.

Space is the other fundamental unit of our existence. One of the things that marks the modern age is a wider understanding of the container space of our universe and of our planet. The incorporation of the wider world into universal and universalizing discourses, although always from specific places and for particular purposes, represents one of the most significant and signal marks of modernism. There are also the more intimate spaces where we lead our more circumscribed lives: the nation-state, the city, the neighborhood. In the Late Modern Wave even the most intimate spaces are shaped by the state, the market, and global connections. Different spatial scales are not separate arenas but different wavelengths in the same continuum.

Space and time are inextricably linked. We occupy time as we inhabit space. The Incas used the term *pacha* to refer to the single dimension of space and time. Physicists use the metric of space–time to build models of our world. There are absolute measures of both space and time, a second is second, and a mile is mile. But these fixed definitions refer to a Newtonian world. In our Einsteinian world we know that time slows at higher speed. Space and time also influence our measurement of both. The Heisenberg Principle tells us that it is difficult to determine simultaneously, with any accuracy, both the location and velocity of particles. Time dilates, spaces warp, and frames of reference matter.

Shallow time and more intimate spaces are socially produced and culturally represented. I will use the term "warping" to refer to the manifold experiences and representations of shallow time and intimate space. I will explore four aspects of this time–space warping.

The first is the creation of universal metrics of space and time. In pre-modern Europe, "measurement was inseparable from the object being measured and the customs of the community which performed the measurement" (Alder 2002, p. 127). There were myriad measures and ways of measuring. One community might measure grain heaped high; another might level it off. Measures had an anthropomorphic bias: the length of a foot, the size of a thumb. They also related to human performance, arable land measured in terms of how long it would take a man to plough a field in a day, for example. Measures were rooted in local customs, tied to human dimensions. This kind of measurement works well in purely local markets and regional economies. But with the development of national and even global economies, standard measures became not only necessary but are an important part of the powerful imaginary prompting wider economic incorporation. The French philosopher Condorcet (1743–1794) shared the Enlightenment project of progress.

He believed in a liberal economy, free public education, and equal rights for men and women. He also conceived a metric system as a powerful way to create a universal language of measurement, freed from the constraints of the local and the weight of the past. The French Revolution allowed more radical pursuits of new measures. For a time, a new calendar was proposed of twelve months of thirty days, and a decimal clock was introduced with the day divided into ten hours of 100 minutes of 100 seconds. For space, a new universal measure was required. The aim was to construct a global metric with the meter to be one ten-millionth of the distance between the pole and the equator. Ken Alder (2002) tells the fascinating story of the two astronomers, Delambre and Méchain, sent out in 1792 to measure a baseline from Barcelona to Dunkirk. They created the modern meter in the process. The dark secret at the heart of the actual measurement provides a thriller-like mystery to Alder's account. He shows that the enterprise, although promoted by notions of rationality and universality, was an all too human achievement.

Just as with measurements of space, so with divisions of time. There used to be such a thing as local time, based on when the sun peaked at noon. Since the sun moves at around 1,100 feet per second, places further apart experience noon at different times. This local time was a more natural time linked to the rhythm of the planet circling around its own axis. To this day, the bell at Oxford's Christ College rings at 9.05 p.m. to signal 9 p.m. local Oxford time. It has done so since the seventeenth century, before the coming of the railways, which undermined the viability of connecting local times. Trains covered space too quickly for local times to be effective. In 1840 the Great Western Railway adopted London time. When it was 9 p.m. in London it would be 9 p.m. in Oxford, no matter when the bell rang at Christ Church. By 1847 all the railway companies in Britain used London time, and by 1855 all public clocks chimed noon in unison. The dominance of London time across the entire country was just one more example of the urban primacy of Britain. Britain is also a relatively narrow country: the distance from east to west limits the time differences to manageable proportions. But even the larger countries were forced to make changes to local time. In the middle of the nineteenth century, there were 144 official times in North America. With the coming of the railway, it became important to have more precise and standardized time keeping. Initially, the railway companies kept their own time. The Pennsylvania railroad maintained Philadelphia time along its entire network. New York Central used the time at Grand Central Station along all of its routes. In St Louis, a major railway junction, there were six official railroad times. It was fitting, then, that in this time-challenged city the managers of fifty large railway companies in 1883 decided to reduce the now fifty different times zones to just four: Eastern, Central, Mountain and Pacific. The new time zones came into effect on Sunday morning, 18 November 1883. Many, but not all, private/public institutions quickly deployed them, and in 1918 Congress ratified the arrangement.

Standard national times provided the necessary basis for a world standard time created in 1884 at a meeting in Washington, DC. The meeting standardized the Greenwich Prime Meridian, the International Date Line and the universal day of twenty-four time zones.

Agreeing upon a universal prime meridian—zero degrees longitude—was a difficult though necessary first step. Previously, states used their own capitals as their prime meridian. The French used Paris, the British used Greenwich in London, the Dutch Amsterdam, the Belgians Brussels and the Portuguese Lisbon. After the American Revolution, mapmakers in the United States changed the prime meridian from London to Philadelphia (the capital from 1790–1800) and then to Washington. Throughout the nineteenth century, many US maps would use a double system on the same page, with longitude from Washington on the bottom and Greenwich at the top. In 1871 an International Congress meeting in Antwerp agreed that sea charts should use an agreed-upon prime meridian. It was to become obligatory within fifteen years. Most countries adopted the new system, although the French still used Paris, the Spanish used Cadiz and the Portuguese Lisbon. The Brazilians used both Greenwich and Rio de Janeiro, and the Swedes, always polite and unwilling to offend, used Stockholm as well as Greenwich and Paris. At the second International Geographical Conference in Rome in 1875, it was agreed to use Greenwich as the prime meridian on land maps, although the conference hoped that this honor would make Britain adopt the metric system. Britain proved recalcitrant on this matter, and feet and yards, stones and pounds, pints and quarts were to last well into the next century. The delegates at the 1884 conference agreed Greenwich would become the prime meridian. The world now had a global metric; local time had been replaced by a standard time centered on Greenwich. Space had triumphed over place. Longitude would be measured east and west of the Greenwich line and continuing the line on the other side of the world yielded the International Date Line, where a new day began when it was noon in Greenwich.

The conference also decided upon a 24-hour clock and the resultant division of the world into twenty-four separate time zones. The initial idea was to have a perfectly even distribution of the earth's surface into equal one-hour units. In reality, as any glance at current maps of time zones reveals, politics bends the straight lines. The International Date Line, for example, bends so that far eastern Russia is not placed in the same day as Alaska. Then there are the anomalies. Australia is divided evenly into three time zones across the country but central Australia is not as the name suggests in the middle of the time zones, but eccentrically located, one-half hour difference from the east coast zone, and one and one half-hour difference from the west coast zone. Just to add to the chronological confusion, one of the states in the eastern zone, Queensland, does not operate daylight savings during the summer months, while the others do. In the central zone, one state uses daylight savings, South Australia, and one does not, Northern Territory. On its own in the western time zone, Western Australia felt free to operate Summer Time, but only from 2006 to 2009. A time zone is also bent from straight so that Iceland is in the same time zone as Greenwich despite being well to the west. Time zones are not only bent, they are also erased. China, a vast country as wide as the four-time zone US, shares just one time zone. When it is noon in Beijing it is also noon in Kashi almost three thousand miles to the west, over three and a half hours later as the sun moves. A Chinese national standard time rules over both local time and world standard time.

The differential adoption of standard time shows the persistence of difference in the face of standardization. The leveling effects of standard time are hinted at in Clark Blaise's assessment:

> Standard time . . . overrode aboriginal time, Hindi and Buddhist, farmer and fisherman crack-of-dawn time, or setting sun Muslim and Jewish time. The standard time begins at midnight in order to avoid the irregular sunrise and sunset of nature. Standard time is a god of predictability and precision. . . . He shows up for work at Greenwich precisely at midnight, every midnight for all eternity.
>
> (Blaise 2000, p. 26)

There were critics of this standardization then as now. In a pre-emptive critique of Friedman's lauding of the flat world and in a very early outing of the notion of Empire discussed by more recent Gallic authors (Deleuze 1995; Hardt and Negri 2000), an early nineteenth-century French writer, Benjamin Constant noted with regard to the metric system: "The conquerors of our times, peoples or princes, want their empire to possess a unified surface over which the arrogant eye of power can wander without encountering any inequality which hurts or limits its view" (quoted in Alder 2002, p. 317).

Cities, especially the larger more connected cities are inextricably networked into universal standard time. Figure 3.2 shows the GaWC global urban network. It is a network meshed into world standard time; Figure 3.3 shows how cities are located in time as well as space, distributed across the 24-hour clock in terms of their financial markets. The world is envisaged as a 24-hour period. The bars with the names of cities represent the hours of operation of their financial markets. The Shanghai stock exchange, for example, is open from 9 a.m. to 3 p.m. local time which is eight hours ahead of Greenwich Mean Time (GMT). New York stock exchange is open from 9.30 a.m. to 4 p.m. local standard which is five hours behind GMT. Other cities are also listed, enmeshed in a standard time that in turn is embodied in the unfolding of time across the space of flows of connected cities. The world is given almost 24-hour coverage through the operation of major centers in the three zones of west, central, and east. While the almost 24-hour coverage requires three major centers the exact location of these centers within these zones is related to history, path dependent developments, and current economic strength. Currently, the big three are New York, London, and Tokyo

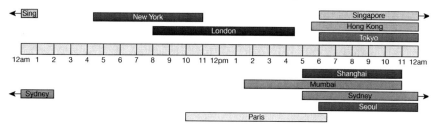

Figure 3.3 Cities in world standard time.

respectively. The diagram also hints at the possible future rivalries. New York's position given the size of the US is secured. In the central time zone London's position seems assured given the existing size and dominance of London's City despite the relative weakness of the UK economy. Berlin would be the most obvious site given the size and resilience of the German economy but the city's development as a global financial center was stunted by the postwar partition of Germany. In the eastern time zone Tokyo is currently the dominant financial center but we can see the rise of other centers especially Hong Kong and Shanghai, which as we will see in Chapter 10 is reglobalizing to its formerly, precommunist role as the Far East's major global financial center. Cities compete in space–time as well across space.

Not only are financial transactions coordinated in time, the global metric allows the space–time sequencing, urbanization and internationalization of events. With an agreed-upon calendar and space–time grid, the world enjoyed a 24-hour celebration of what many considered the dawn of the new millennium on January 1, 2000. Here I draw upon Short (2001). First up was Kiribati in the Pacific. This nation of thirty-three islands, of coral sand and rock fragments that 85,000 people call home, was the first country to see in the new millennium at midnight local time, 12.00 Greenwich Mean Time. Three years earlier Kiribati, in preparation for the event, had the date line moved to follow the country's eastern border. The nation had also joined the United Nations in September 1999, just in time to join the international community before the big day. Kiribati wanted to have the honor and media attention of being the first nation into the new millennium. Hour by hour the new millennium passed around the planet. Watching at home on television, I was never more aware how much the earth was like a giant clock with local places all part of the same shared space and common time. An hour later, fireworks in New Zealand. Two hours later Sydney put on a marvelous show with fireworks that lit up the bridge and Opera House in a wild display of pyrotechnics. The animated image on my television screen showed the line moving steadily westwards. Tokyo and Hong Kong now. Most Asian countries traditionally used a lunar cycle. Under the Chinese traditional calendar, which works on a sixty-year cycle, the year 2000 was lunar year 4698. The New Year would traditionally fall on the second new moon after the winter solstice—anytime between January 21 and February 19. In Korea, Japan and China the millennium celebrations revealed a dual time, traditional and world standard, and the growing importance of the global standard. Not everywhere celebrated. In Pakistan and Afghanistan the official authorities did not celebrate what they considered a foreign festival. Many Muslim fundamentalists consciously sought to distance themselves from this global event and in Dhaka, the capital of Bangladesh, 500 police were called in to stop revelers from drinking alcohol. And in Israel rabbis banned New Year celebrations. The silences revealed much about the resistances to globalization.

The line kept moving. Rome. Berlin. Then there was Paris. The Eiffel Tower was so ablaze with fireworks I thought it was going to take off. The French had something to celebrate: not only did their celebrations go really well, but also the English seemed to have fluffed theirs. The giant Ferris wheel in London did not work, the river of fire never materialized, and watching the Queen hold hands with commoners was like watching a very old distant aunt, trying to fake jollity at a family event with people she really did not like.

The high point in time for me, like everyone else I assume, was when the New Year came to my neighborhood. Celebrating with friends, drinking champagne, and watching the revelry in Times Square. Midnight passed on to the Midwest, the mountain zone, and then the west coast. The line kept moving until French Polynesia, one of the last places on earth to see in the New Year.

The millennium celebrations show us the power and significance of global spectacles. They make us realize the shared yet different experience as people around the world celebrate the same thing at different times. They also reveal that global events are concentrated in the big world cities around the world, the places that lit up as midnight struck; the rural parts of the same country remained darker, less connected to the global celebration.

The second element of space–time warping, closely connected to the first, is space–time convergence. It takes time to cover distance. To overcome space takes time. Space–time convergence occurs when the time taken to cover distances shrinks.

Convergence is closely associated with waves of globalization. The second wave of globalization initiated in the late nineteenth century was associated with major space–time convergences including the railways, the telegraph, and the steamship. It took less time to travel large distances. People, ideas, and messages could move more quickly. News of events was more quickly transmitted across the globe, in effect shrinking the globe to a more compressed space. Space–time convergence effectively shrinks the globe as people and events are brought closer together. The world is flattened as it is shrunk. The physical changes also impacted cultural representations and scientific understandings. The railway station, Gare Saint-Lazare in Paris was the setting for Edouard Manet's 1874 painting *The Railway*, which recorded the new urban landscape. In the painting a woman sits and a child stands, the woman looking toward the viewer while the young child looks away through an iron fence at a grey trail of smoke. The title seems enigmatic. There is no railway visible. The railway engine has left a transient trail of steam, it has gone, the space that it only recently filled now filled with the visible presence of its absence. Space and time subtly and whimsically counterpoised. The space–time convergences of this wave of globalization are reflected in the cultural and scientific representations of space and time as fractured, plastic, and relational. The experience of the new technologies in the period from 1880 to the Great Depression, what I have termed the second wave of globalization, also influence how people saw the world, experienced the world, and represented the world (Kern 1983). Convergences create new attitudes to space and time, a new awareness of space–time's twisting and fracturing. Picasso's Cubist representation of space and Einstein's new models of space–time occur in this period. The modernist novelist James Joyce worked on his masterpiece *Ulysses* during the nineteen teens. He began it in 1914. Parts were serialized in a literary magazine in 1918. It was completed in 1921 and first published in 1922. In the novel, set in Dublin, each of the eighteen chapters covers approximately one hour from eight in the morning until two in the morning. This rigid external time schema is counterpoised to the time warping streams of consciousness of the characters' internal thoughts. Virginia Woolf's 1925 novel, *Mrs Dalloway*, deals with the external events in one day but also explores the thoughts in the minds of two main characters where time stretches backwards and forwards as they reminisce and remember. One character, preparing for a birthday party, remembers her marriage choices, another suffering from trauma from remembered experiences in World War I spends the same day in a park. Later that evening while the birthday party takes place, the veteran commits suicide by jumping out of a window. The time fractured sense of the novel was given a contemporary twist in the 1999 novel *The Hours* by Michael Cunningham and the subsequent 2002 film adaption. This late modernist version of the modernist classic includes a portrait of Woolf creating the novel, a late modern self-consciousness about the creative process layered onto the early modernist sensibilities.

Each wave of globalization is associated with new forms of space–time convergence and the resultant change in their representation. David Harvey (1989b) makes the case that new cultural forms arise around 1972, just about the time of the speeding up of what I term the Late Modern Wave of Gobalization. He connects postmodernism ultimately to the new flexible modes of capital accumulation and new rounds of space–time convergence. In each major wave then, space–time convergence brought about by changes in technology and the reorganizations of economies, has consequences for how we experience, manage, and represent space and time. In the Late Modern Wave, issues of global connections and flows are more central. Compare the early modern classics *Ulysses* and *Mrs Dalloway*, both set in one city, with a late modern representation such as the 2006 film *Babel* with its multiple and interconnected stories unfolding in Morocco, Japan, Mexico, and the United States. Border crossings, international flows of people and goods all presented in non-linear space–time sequencing.

Cities are the pivotal points in space–time convergences. They are the hubs in the space of flows. The late nineteenth-century railway station has as its contemporary locale, the international airport. But more than just housing the means of shrinking space, cities become the embodiment of a shrinking space. They generate, consume and transform the flows of goods, people, and ideas.

A third dimension of space–time warping is space–time compression. If convergence shrinks space then compression forces more things into the same time unit. The backdrop is the increasing global competition and connectivity that leads to a greater productivity. And productivity is in effect a collapsing of the time taken to do the same unit of work. A flatter world requires more things than previously to be done in the same time unit. The increasing circulation time of capital, the lowered horizon for profitability, and the wider competition for any economic activity, all conspire to make work accelerate. This is not just a macroeconomic effect. The associated technological changes in our personal lives include the cellphone, email, texting, and the like. We are expected to do more work, to cover more space, in the same or even less time. A paradox emerges whereby the new technologies that promised us vast leisure time (and that was the early promise of the computer) have created time shortages. The amount of work—and work can be defined very broadly as things done either in public or private, in the household or for an employer —has increased, giving most of us the sharp sense that we neither have enough time nor the time necessary to accomplish all that we are required to do or feel that we need to do. Time, which was gained in space–time convergence, is lost in space–time compression as we struggle to keep up with the demands to do more work in shrinking amounts of time. The space we can cover expands while the time we have to do all we need to do shrinks. New technologies promise time liberation but deliver time bondage.

Cities, finely calibrated to the universal standard time, are places where this compression leaves the heaviest imprint. It is in the cities where people are texting and twittering, emailing, and Facebooking in electronic space while they try to navigate the physical space of the metro region as quickly as possible. In places of dense concentration of people, the space–time compression maximizes the amount of things to be done into shorter intervals of time. Things accelerate in the city.

A fourth element is what I will term space–time splicing. This is the juxtaposition of times and spaces. Using such elementary categories as *then* and *now* for time and *here* and *there* for space we can fashion a rudimentary grid. *Here and now* is the perpetual present permanently poised between the past and the future. To live in the now is the teaching of many spiritual leaders, which gains extra traction during periods of extended space–time compression. It is no accident that the practice of yoga, which emphasizes living in the here and now in both body and mind, is increasing in popularity for the time-oppressed classes of the major cities.

Here and then is the production of space and place with the syntax of the past. Postmodern architecture, for example, jettisoned the flat roofs, straight lines, and no ornamentation of high modernism and instead opted for pediments, decoration, historical referencing, color, pastiche, and whimsy. If Mies van der Rohe opined that less is more, the postmodernist architects replied with less is bore. The archetypal city of the *here and then* is Las Vegas with its fake Eiffel Tower and phony Venetian canals all built in the Nevada desert. Difficult, uncertain, and unsettling times create demand for the phantasmagoria of the past: reimagined and manufactured pasts that tell us more about the uncertainties of the present than any historical "reality."

In globalizing cities around the world, there is a building boom that features architectural syntaxes of both the *here and now* and the *here and then.* The *here and now* of global connectivity and global competitiveness is visually rendered in the new airport, the shiny corporate headquarters, and the cultural center by the signature architect. The architectural *here and then* embraces late modernism's restored historic areas, ornamental quotations from the past, and a self-conscious creation of historical depth.

Simply to be modern is to be dated. What is required is a blend of the economic efficiency of the functionally modern with the cultural sophistication of the reworking of the past. The built form in cities around the world displays both the *here and now* as well as the *here and then.*

There and now is the condition of a compressed and flatter world. News of an earthquake in Chile, devastation in Haiti, the war in Afghanistan, or the fall of a currency are transmitted instantly around the world. It is not only simple news that is quickly distributed; it is also a global awareness and a global connectivity. Cities are the hubs of this network not only as distribution points but also as epistemic centers where information is turned into knowledge and data is transformed into the basis for strategies. Cities are ensembles of information processing and knowledge production.

There and then is the discourse of understandings of the historical experiences of places other than our own. Knowledge of others' history is partly dependent on geography. Edward Said's work, for example, drew attention to the invention of the Orient by those in the West. The West is also portrayed by those in the East. There is an Occidentalism as well as an Orientalism. The West is also understood and imagined by its enemies. The US, for example, is often projected as a "rootless, cosmopolitan, superficial, trivial, materialistic, racially mixed, fashion-addicted civilization" (Buruma and Margalit 2004, p. 8).

There is no shared *there and then* narrative even in the current superpower, the US, the main source of the flat/smooth space of Empire. A striking feature of life in contemporary America is the enormous ignorance of the outside world. *There and then* is poorly understood by many Americans. This is not a new state of affairs. Tocqueville, almost 200 years ago, notes that the American people are blissfully ignorant and geographically isolated. There is knowledge of foreign countries and places in specialized institutes, universities, and colleges. This knowledge is concentrated in major cities. While New York is attuned to global financial changes, Washington, DC is sensitive to political fluctuations, and Los Angeles listens very carefully to the changing preferences of a global movie audience. However, the educational system, especially at the elementary and high school level, spends little time on the history and geography of other parts of the world. The emphasis is on the *here and now* of contemporary US compared to the *there and then* of other parts of the globe. The vast bulk of pre-college education in the United States is concerned with events in the US. Most educational systems have a national bias; they are after all a crucial element in the creation of national identity and consciousness. Schools and universities, as Louis Althusser reminded us, are part of the ideological state apparatus, although this does not mean that they cannot also act as sites of resistance to national ideologies. The US insularity is reinforced by a number of factors. In many other countries the simple facts of geography and the disputed facts of history have created a spatial awareness of others. Sharing borders with only two countries, separated by two oceans, from Europe and Africa on the one side and Asia from the other, the US is geographically less connected to the rest of the world. The sheer size of the country also means that internal travel provides continental difference. Unlike sun seekers in northern Europe, people in the US do not need to leave their own country to experience tropical climates or desert landscapes. The historiography of the US has been dominated by internal affairs. The bias of hereness in popular understanding in the US of the rest of the world is reinforced by the political power of the US. At a military level, the US is the undisputed superpower, perhaps the only one with sustained global reach and capabilities. The US projects power on others. The *theres* have to be aware of the *now* of the US. There is two-way intellectual traffic but the export rather than the import trade dominates it. The US is such a powerful presence in the world that people in other countries have to learn about the US. They are made acutely aware every time they hear world news or watch a Hollywood movie that there are other places in the world with a different history. The rest of the world is continually given an education in the *there and now* and *there and then*.

Spectacle, performance and global urbanity

Urbanization is unfolding against a background of universal standard time, space–time convergence, space–time compression and various forms of space–time splicing. The unfolding relationships give character to the nature of modernity and guide the arc of its development. Let us now consider just one of the many connections.

Figure 3.4 Global hub: interior of Paris airport.
Photo: John Rennie Short.

Guy Debord coined the phrase "society of the spectacle," asserting, the spectacle is the chief product of present-day society that is increasingly capitalist and global (Debord 1973). In the society of the spectacle social relationships are mediated in and through images. I will extend the definition of images beyond the two-dimensional to the three dimensions of built form and the space–time dimension of performance and enactment. A metropolitan modernity is enacted in the creation of the global hubs that include international airports, business centers, hotel districts, entertainment areas, and market residential areas that cater to the cosmopolitans; in the production of symbolic buildings; and in the space–time performances or urban global spectacles (see Figure 3.4).

Global hubs are geared toward connections—not just the travel and communication networks, but also the discursive networks of the latest trends. A mundane example but an important one: the experience of a business-class hotel now has a global minimum standard. Rooms are equipped with fluffy towels, bathrobes, and expensive shampoos. The result is to create a minimum level of service in order to be accepted into the category of an international business-class hotel and a smoothing of the business-class hotel experience in cities across the world. Sometimes the semiotics are confused. I once spent two nights in a business hotel in Pudong in Shanghai, where there were the requisite fluffy towels. But all of them were bright pink.

The presence of the coffee house and sushi bar are just some of the many signs that cosmopolitanism is at work. Cosmopolitan sites are a mark of global

Figure 3.5 Corporate headquarters: IAC Headquarters NYC, designed by Frank
Gehry.

Photo: John Rennie Short.

modernity. Often associated with these sites is the work of signature global architects. One of the most recognizable global languages is architecture, a commercial art form that turns visions into concrete realities and solidifies messages of globality, power, and prestige. The term "signature architect"— or "starchitect"—is used to refer to a small group of well-known architects whose very names have a natural aura of architectural prestige. They are hired to design the set pieces of architectural spectacle—the global hubs, the prestigious corporate offices, and the cultural ensembles. Examples of the architectural hubs include the air terminal in Osaka designed by Renzo Piano and Eero Saarinen's design for Dulles airport. There are the corporate headquarters: I. M. Pei designed the Bank of China building in Hong Kong, Frank Gehry designed the IAC office in New York City, and Norman Foster did the Commerzbank headquarters in Frankfurt and the Hong Kong and Shanghai Bank in Hong Kong. There are also the cultural ensembles: Joern Utzon's Sydney Opera House, CalI. M. Pei's Meyerson Symphony Center in Dallas, as well as the East Building of the National Gallery of Art in Washington, DC, James Stirling's Art Gallery in Stuttgart, Frank Gehry's Guggenheim Museum in Bilbao, Renzo Piano's Pompidou Center in Paris, Santiago Calatrava's Art Gallery in Milwaukee, and the Public Library in Seattle designed by Rem Koolhaas (see Figures 3.5 and 3.6). The new Reichstag in Berlin was designed by Norman Foster. Signature architects are part of the globalizing projects of cities. In Los Angeles, long considered culturally provincial by the eastern elite in the US, there has been a self-conscious use of big name architects to produce sites of conspicuous culture. The Getty Center designed by Richard Meir opened in 1998, and the Walt Disney concert hall, designed by Frank Gehry opened in late 2003. The spectacular building by the signature architect is now part of the stage setting for an urban globality.

The performance, enactment, and witness of metropolitan modernity fortifies urban globality. In Chapter 4, I will show how the global spectacles

Figure 3.6 Cultural ensemble: art gallery in Milwaukee designed by Santiago Calatrava.

Photo: John Rennie Short.

of World's Fairs and Olympic Games played and continue to play an important role in the performance of modernity on a metropolitan stage. For the moment I will make some general remarks about the Olympics as an example of the relationship between cities, modernity, and global spectacle.

The first games of the modern era, held in Athens in 1896, involved 241 athletes from only fourteen countries and limited press coverage. Over the years, the Games have grown in size, scope and international media coverage. Over 11,000 athletes from over 200 countries participated in the 2008 Games. The Games are now the most watched events on television, with a truly global audience. The Olympic Games embody the increasing globalization of the world: they represent a significant regime of international regulation, embody a shared cultural experience and provide an important platform for economic globalization as transnational corporations advertise in and through the Games. Hosted by cities, the increasingly global Games are worldwide events that unfold in a particular place. The Olympic city plays host to the world, "theatricalizing" the city and making it a media spectacle unto itself. As a vehicle for urban representation and landscape alteration, the Olympics and similar events contribute in various ways "to a profound shift in our relations to our urban spaces, spectacularizing them in the interests of global flows" (Wilson 1996, p. 617).

Cities' elites are involved in the maintaining, securing and increasing of urban economic competitiveness in a global world. This involves improving global connections, and enhancing global visibility as signposted by cosmopolitan hubs, the work of signature architects and a metropolitan modernity enacted in global spectaculars.

CASE STUDY 3.1 **Seoul revisited**

The first time I visited Seoul was in 1993. On subsequent trips I noticed a number of changes. Some obvious. The banks of the Han River are much greener, more landscaped and more accessible. The city as a whole looks much greener than the uniformly grey of the concrete city I remember. Air quality, despite the huge increase in traffic, is noticeably better although still not great. The sheer weight of traffic as well as the airborne pollution from China creates a pall of smog across the city. Seoul is more globalized: there are Starbucks and more signs in English. South Koreans are connected to the rest of the world through economic linkages as well as diasporic connections and internet traffic.

I visited a city that was the epicenter of spectacular change and rapid modernization. The city's population in 1949 was around 2.4 million before it embarked on decades of sustained economic growth fueled by export-led manufacturing and population growth based on rural to urban migration within South Korea. In the 1960s and 1970s the city was increasing in population at the rate of half a million every two years. The enforced industrialization/urbanization led to major environmental damage and social dislocation but did lift the majority of South Koreans out of poverty. Authoritarian governments ensured that breakneck economic development was not halted by democratic discussion. Massive urban renewal involved clearance of squatter settlements, slums and red light districts, roads were widened, and the Han River was channeled in high embankments. Seoul was the center of rapid urbanization. The nation's population classified as urban increased from 14.5 percent in 1950 to 88.3 percent in 2000. By 2010 the city population was 10.3 million. Despite the growth of other cities in South Korea, Seoul retains its national prominence, dominating the national economy, politics, art, and culture.

There was a rapid rise and relative fall in manufacturing employment in the city that peaked at 1.3 million in 1990, 30 percent of the labor force. By 2000 manufacturing constituted only 19 percent of total employment. The city's economy is now dominated by the service sector employing 3.6 million people, 80 percent of the city's total workforce. The city has moved from a manufacturing to a service center. It also spread out as urban growth leapfrogs the green belts with the construction of new towns and the widespread construction of high-rise towers turns a concentrated city into a polycentric metropolitan region (Kim 2003). The city moved from high-density city to a more dispersed metropolitan region tied together by mass transit and increasing car usage. The number of registered vehicles increased from 206,000 in 1980 to 2.2 million in 2000; although that was very easy to believe as I sat with my friend in traffic jam after traffic jam as we tried to make our way across the city.

The city has also globalized. The city and the nation used the hosting of the 1988 Summer Olympics to create a more positive global image. In preparation subways were built and part of the central business district was rebuilt.

Seoul is South Korea's most important global hub. However, the ethnic homogeneity of the city continues to be very obvious compared to other powerful economies. With less than one percent foreign born, the city's ethnic and racial homogeneity is maintained by strong controls on foreign immigration. Seoul is more economically than culturally globalized. Yet the pursuit of global city status dominates national and city policies. The catchphrase of the thirty-third mayor Oh Se-hoon was "A Clean, Attractive and Global City" and he promotes global zones in the city where foreigners have fewer language problems. However, Kim and Short (2008, p. 172) note that "It remains to be seen whether the government will place cosmopolitanism, multiculturalism, openness, tolerance, and social inclusion at the center of any future strategies for cementing Seoul's world city status."

Over the course of my visits I also noticed more subtle changes. A more pragmatic attitude to the costs as well as benefits of rapid economic growth is increasingly evident from speaking with the people and listening to their debates. There is a fuller reckoning of the colonial legacy of Japanese control, of the collaborations as well as the resistances. While there is justifiable pride in economic achievements, there is a deep sense of the trauma inflicted on the country by the recent history of colonialism, war, and rapid industrialization. There is a palpable grappling with history. Seoul now boasts a new National Museum of Korea and the grand Joseon palaces of Changdeokgung and Gyeongbokgung are refurbished and open to the public.

Other visible reminders of the past are not such venerable legacies. You do not have to travel very far from Seoul to see the barbed wire that separates South and North, a developed country from one bordering on mass starvation. Since my last visit, South Korea has prospered while North Korea has spiraled down into even more authoritarian rule, misery, and hunger.

South Korea seems poised to enter a more settled period in which the legacies of the past and the prospects of the future are now regarded as more nuanced and complex. To visit the fast-paced metropolis of Seoul is to get a glimpse of the speed and feel of a high-octane modernity coming to terms with its past and establishing its future role in the world.

———————

Part 2

Thematic specificities

4   Urban spectaculars

World's Fairs and Summer Olympics

Globalization and modernity are constructed and maintained in many ways. One of the most important is through the experience of mega-events and spectaculars hosted in particular cities. The circulation of these events in different cities across the world both creates and tightens global urban networks. These spectaculars connect cities and societies through global discourses and shared practices. The host cities also have opportunities for achieving or reaffirming global city status, acting as platforms for globalizing trends, and laboratories for future urban forms. The host cities are hubs and exchanges in global flows and networks, transmission points in the production of a global society (Gold and Gold 2005; Roche 2006).

In this chapter I will consider two of the most important mega-events of the modern period, the urban spectaculars of World's Fairs and Olympic Games. World's Fairs were vital elements in the globalization of modernity in the period from 1850 to around 1940. World's Fairs gave meaning to modernity: much of today's modern culture has roots in World's Fairs. Olympic Games, in contrast, represent an important staging of the contemporary wave of globalization and late modernity, especially from around 1960 to the present day.

World's Fairs

An important date in the history of the relationship between globalization and cities is May 1, 1851. On that day the Great Exhibition of the Works of Industry of all Nations (hereafter Great Exhibition) opened in London's Hyde Park. The main building was the Crystal Palace, designed by Joseph Paxton. An innovative structure combining glass greenhouse architecture and arcade passages, the Crystal Palace was realized in record time through a modular construction system that involved shaping huge sheets of glass around an elegant framework of prefabricated iron sections. The effect was a light-filled building open to the sky yet closed to the elements. It was a supremely self-confident design, befitting the highpoint of British global prominence, an incorporation of nature that also signified human control over nature. Inside the imposing pavilion—563 meters long and 138 meters wide—a total of

13,000 exhibits were displayed. The exhibits came from around the world, as countries displayed their technological advancements. The US exhibit, for example, displayed the Colt revolver, the McCormick reaping machine, and the Singer sewing machine. The majority of exhibits originated in Britain and reflected the industrial might of the world's largest trading nation and imperial power. Inside the glass cathedral the latest designs and appliances of modern industrial capitalism were brought together under one roof, displayed, and worshipped.

Contemporaries were amazed at the range of sheer range exhibits. Here is Charles Dickens:

> The railway engines, and agricultural engines, and machines; the locomotives, in all their variety; the farm-engines, such as the compound plough, the harrow, the clod-crusher, the revolving sub-soiler ... the draining-plough, the centrifugal pump, the sowing-machine, the reaping, the thrashing, and the winnowing machines, the chaff-cutter, the barley-hummeller, the straw-shaker, the combined thrashing, shaking, and blowing machine ... sawing-machines of great power; machines for planing; others by which a large hurdle can be cut from the solid timber, and put together in nine minutes, and a fifty-six gallon beer-barrel made in five minutes.
>
> (Dickens and Horne 1851, p. 358)

Such are the new-fangled effects of the tools of space–time convergence and space–time compression. But not all contemporaries viewed it with such slack-jawed wonder. Karl Marx, for example, saw it as the ultimate fetishization of the commodity form. A more common response was a wonder at the size, diversity, and global reach. After her visit, Charlotte Brontë remarked:

> it is a wonderful place—vast, strange, new and impossible to describe. Its grandeur does not consist in *one* thing, but in the unique assemblage of *all* things. Whatever human industry has created you find there, from the great compartments filled with railway engines and boilers, with mill machinery in full work, with splendid carriages of all kinds, with harness of every description, to the glass-covered and velvet-spread stands loaded with the most gorgeous work of the goldsmith and silversmith, and the carefully guarded caskets full of real diamonds and pearls worth hundreds of thousands of pounds. It may be called a bazaar or a fair, but it is such a bazaar or fair as Eastern genii might have created. It seems as if only magic could have gathered this mass of wealth from all the ends of the earth—as if none but supernatural hands could have arranged it thus, with such a blaze and contrast of colours and marvellous power of effect.
>
> (Shorter 1908, p. 46)

The poet Tennyson referred to the 1851 Great Exhibition as the "world's great fair" and the name stuck.

The connection between the exhibition, space–time convergence and a capitalist globalization was apparent in Prince Albert's speech, delivered two years before the exhibition opened, when he noted:

> The distances which separated the different nations and parts of the globe are gradually vanishing before the achievements of modern invention, and we can traverse them with incredible ease; the languages of all nations are known and their acquirements placed within the reach of everybody; thought is communicated with the rapidity and even by the power of lightning.
>
> (Martin 1880, p. 247)

While there had been World's Fairs before 1851—the first was held in London in 1756 and awarded prizes for improvements in the manufacture of tapestry, carpets, and porcelain—the Great Exhibition marked the formal beginning of the modern World's Fair and its association with the globalization of a capitalist modernity. Over six million people visited the Exhibition, and it was widely considered the leading edge of industrial capitalism that was sweeping across the world. The Great Exhibition was so successful that it was widely copied with intent to surpass (particularly by France). Other cities in many other countries wanted to host World's Fairs. The sheer number is impressive, a total of 265 from 1851 to 2010. The golden era of World's Fairs was from 1870 to the 1940s, during a pronounced wave of globalization, when over twenty Fairs were held each decade. While the large urban centers such as London and Paris hosted fairs, some multiple times, other smaller cities from across the globe also hosted fairs, including British colonial cities such as Cape Town, Brisbane, Melbourne, and Sydney. Host cities were scattered throughout Europe, the Americas and Asia, and included Tokyo, Shanghai, and Nanking. There were so many World's Fairs during this golden era because it was a time of rapid globalization. And rapid globalization, in turn, was aided by the World's Fairs.

Hosting a World's Fair signaled economic achievement, raised the profile of the city, leveraged national public funds for civic purposes and influenced global public opinion. Indeed World's Fairs were part of the construction of global public opinion. Global city status was attempted and sometimes confirmed by and reflected in hosting a major World's Fair. The Bureau of International Expositions (BIE) sanctioned a small number of World's Fairs. An international organization established in 1928 originally consisting of thirty-one countries, the BIE now comprises 156 member countries. Table 4.1 lists the sanctioned World's Fairs, some retrospectively. This list is dominated by, especially in the earlier years, London and Paris, as the two capitals of rival imperial powers compete with each other. It is not until 1970 that an Asian city hosts a BIE sanctioned event. There is an even smaller list of fairs considered so large and important that they significantly influenced global trends enough to be considered landmark fairs (Findling and Pelle 2008). These are also noted in Table 4.1.

Table 4.1 World's Fairs 1851–2010

1851	London, UK*
1855	Paris, France*
1862	London, UK*
1867	Paris, France*
1873	Vienna, Austria-Hungary*
1876	Philadelphia, US*
1878	Paris, France*
1879	Sydney, New South Wales
1880	Melbourne, Victoria
1888	Melbourne, Victoria
1888	Barcelona, Spain
1889	Paris, France*
1893	Chicago, US*
1897	Brussels, Belgium
1900	Paris, France*
1904	St Louis, US*
1905	Liège, Belgium
1906	Milan, Italy
1909	Seattle, US
1910	Brussels, Belgium
1913	Ghent, Belgium
1915	San Francisco, US*
1929	Barcelona, Spain
1933	Chicago, Illinois, US
1935	Brussels, Belgium
1937	Paris, France*
1939	New York, US*
1949	Port-au-Prince, Haiti
1958	Brussels, Belgium*
1962	Seattle, US
1964	New York, US
1967	Montreal, Canada*
1968	San Antonio, US
1970	Osaka, Japan*
1974	Spokane, US
1975	Okinawa, Japan
1982	Knoxville, US
1984	New Orleans, US
1985	Tsukuba, Japan
1986	Vancouver, Canada
1988	Brisbane, Australia
1992	Seville, Spain*
1993	Daejeon (Taejon), South Korea
1998	Lisbon, Portugal
2000	Hanover, Germany
2005	Aichi, Japan
2008	Zaragoza, Spain
2010	Shanghai, China*

Note The list only includes fairs sanctioned by the BIE.
* indicates landmark fairs.

World's Fairs embodied a particular form of capitalist modernity and displayed a global world of linked national economies. While each fair promoted this general message, they also each had a touch of particularity. The Great Exhibition, for example, gave the promise of a new world, one where human ingenuity could conquer the limitations of nature. It was also a socially encoded promise that assumed a complex set of ideas, including British industrial might, free trade, pacifism and what was called universalism. The very form of the exhibition, in which national products were under one roof, showed how national economies compared and contrasted but also were linked and connected. Subsequent fairs had complex, often conflicted, messages as nationalism competed with free market ideologies, national discourses were often fractured between competing elites, and government power had to deal with consumer choice.

World's Fairs signified a connected world in a global urban circuitry of regular events through which flowed people, ideas, and capital. One dominant idea was the importance of technological innovations and inventions. World's Fairs introduced many new goods to large audiences: the Colt revolver and the McCormick reaper (London, 1851); the elevator (Dublin, 1853); the sewing machine (Paris, 1855); the calculating machine (London, 1862); the telephone (Philadelphia, 1876); outdoor electric lighting (Paris, 1878); the gas-powered auto (Paris, 1889); motion pictures (Paris, 1900); controlled flight, the wireless telegraph (St Louis, 1904); Kodachrome photos, (San Francisco, 1915); television (New York, 1939); computer technology, fax machines (New York, 1964). For much of their history Fairs presented futures of technological wonder and a technical fix to social problems. Human ingenuity could overcome not only distance and disease but also want and hunger. Medical technologies such as the incubator and the X-ray machine were first exhibited at World's Fairs. The Fairs became sites of technology transfer and adoption wrapped in an ideology of technological triumphalism. However, more recent World's Fairs provide something of a corrective to this easy and glib technological optimism. The theme of Brussels World's Fair of 1958 was a *More Human World*, while the theme of the 2005 Expo in Aichi in Japan was *Nature's Wisdom*. The theme of Expo Shanghai 2010 was *Better City, Better Life* and focused on ecological issues. While technological accomplishments still figure largely, there has been a more palpable greening of attitudes and a less naïve sense of what technology, on its own, can accomplish. A heady modernity, which asserted ownership and control of the future, has been replaced by more sophisticated late or second modernity with a less breezy attitude to the future.

A global discourse was created by the large World's Fairs. Let us just consider the early Fairs held in Paris. At the 1855 Paris Fair, for example, the Bordeaux wine classification still in use today was first introduced. At the 1867 Fair, visited by more than nine million people, the Japanese woodcut prints on display helped to shape the modern art movement of Impressionism. At a meeting held alongside the 1878 Fair, international copyright laws were first formulated. The main symbol of the 1889 Fair was the Eiffel Tower, which became an icon not only of the city of Paris in particular but of contemporary urban design. The Eiffel Tower is one of the best examples of modern urban placemaking linked to a particular urban structure (Thompson 2000). The Eiffel Tower is just one of the enduring legacies of both the actual and imaginative geography of a host city that through its hosting solidified its world city status and gave to the world a symbol of both the city in particular and of urban modernity in general. Javanese music played at this Fair also influenced the development of atonal modernist music. More than fifty million attended the 1900 Fair where talking films and escalators were first shown to the public.

The Fairs were also sites of representation of the wider social world and its alterity. Other races and nationalities were presented often in the discourse of imperial and colonial justification. Fairs served as windows on the world in which various messages were coded—material progress, technological triumphalism, national cohesion, white supremacy, "noble savagery" grateful for modernization and thus justification of empire—as well as decoded—adherence to materialism, belief, and reliance on technical progress to solve social ills, mass consumption of imperialism, the "rightful" hegemony of capitalism. Fairs promoted and justified as they explained and narrated the modern world.

The Fairs were not just sites of technology transfer; they were hubs in the global transmission of cultural practices and worldviews that communicated the aesthetics of modernism. There are connections between the development of modern art and architecture and World's Fairs. At the 1929 Barcelona Fair, Mies van der Rohe designed an early example of the International Style. At the 1937 Paris Fair, Picasso's Cubist *Guernica* was displayed, and many of the pavilions were built in a modernist style. Le Corbusier designed buildings for the 1925 and 1937 Paris Fairs. The World's Fair of 1939 in New York, for example, is considered a seminal moment in the maturing of an international modernist movement. Alvar Aalto designed the Finnish pavilion. The Fair, the official motto of which was *World of Tomorrow*, presented a model of mass-market modernism (Christie 2006). Then there were the more incidental connections. For the 1964 New York World's Fair, Andy Warhol's *Thirteen Most Wanted Men* mural on the outside of the New York State pavilion featured silk-screened mugshots of the FBI's most wanted criminals. The New York Governor, Nelson Rockefeller, ordered the piece removed. The censorship added to Warhol's fame in the New York art scene and indirectly helped to promote Pop Art.

In their attempts to display and narrate modernity, World's Fairs shaped and reshaped ideas about class, race, gender, and nationality as well as about progress, economic development, architectural practice, and urban living. They were also used to change global and national optics. Countries represented themselves at Fairs in an attractive light to industrialists and investors. World's Fairs were important stages to sell a country or a region. The state of Mississippi had a stand at World's Fairs from 1884 to 1904 that included African Americans in such a way as to try to dispel the racist image held by many of the deep southern states. At the World's Fairs held in Atlanta (1895), Nashville (1897), and Charleston (1901/2) the host cities sought to present the image of a new modern South free from the rural backwardness of the past. These World's Fairs embraced as well as promoted modernity. The contradictions of modernity were also embodied. Harvey (1996) explores Fairs as sites where nation-states and multinational corporations compete in representational space as they do in the global economy. Pred (1995) exposes the contradictions in World's Fairs as representing the three phases of modernity: industrial modernity, high modernity and hypermodernity. World's Fairs not only reflect a globalizing world, they help to create it.

The Fairs, especially the large Fairs, also gave a glimpse of new forms of urban living, new templates for global and globalizing cities. The new metropolitan spaces depicted at the Fairs were managed and controlled; they gave order, discipline, shape, and form in contrast to the chaos of the cities around them. From the Great Exhibition onwards, Fairs introduced the notion of the managed urban crowd. Urban tourism was initiated by Thomas Cook, as were forms of mass eating and drinking for the huge crowds, in the selling of the Exhibition to foreign and national visitors. World's Fairs introduced night illumination, mass transit, escalators and air conditioning: they "imagined" cities given over to consumption and entertainment spaces.

Fairs in the US in the 1930s, for example, consistently explored the theme of technological solutions, such as air conditioning, to achieve a "utopian" urban life, free of the vagaries of weather and climate. A technological urban utopia was imagined.

Modernism looks to the future and is an especially attractive proposition when the present looks bleak. There were six World's Fairs held in the US during the Depression years of the 1930s (Rydell and Schiavo 2010). It began with a Fair in Chicago in 1933–1934 attended by between 39 and 49 million people, through San Diego 1935–1936, Dallas 1936, Cleveland 1936–1937, San Francisco 1930–1940, ending with the New York World's Fair in 1939–1940 attended by over forty-five million people. The simple attendance figures undercount the impact of the Fairs as their ideas and images were spread via radio, newspapers and magazines. It is estimated that over 220 million people saw the newsreels of the New York Fair. The Fairs occurring during a huge economic downturn displayed a confidence in the future. "The Depression is over," noted the official program to the San Diego Fair, "our faces are forward." A brighter future was presented in an extravagant display of modernism that promised abundance, prosperity, and affluence; new cities lit by electricity and new neighborhoods of model homes filled with streamlined designs for toasters, vacuum cleaners, and dishwashers. A better future was presented and promised by corporations selling not only their products but also a belief in capitalist consumerism. Home ownership and unrivalled individual consumption were promoted in a world of sleek designs, electric light, glass, and steel encased modernist architecture. General Electric, Chrysler, and other major corporations invested heavily in presenting their products in large exhibits. Pan Am suggested a world opened up by airline travel and Ford spent over $12 million at the six Fairs of the 1930s presenting their cars. These World's Fairs, at a time of mass unemployment and a crisis of confidence in capitalism, suggested that the big corporations could do the job of delivering a better future. The progressive elements of European modernism were more shaped in the US by a corporate capitalism. At the World's Fair in 1939 in New York, a 35,000 square foot exhibit space entitled Futurama, funded by General Motors, showed a city of high-rise towers and fast, free-flowing motorways filled with private automobiles; it was a tantalizing vision presented to almost five million visitors. The plan unveiled at the Fair, which drew upon the ideas of urban modernism of the early twentieth century, was the blueprint for urban planning in postwar urban America.

As new urban forms were distilled from the host cities, a recognizable metric of modernity was established. Cities now required mass illumination, modern transport, and modern technological solutions in order to be considered modern. World's Fairs displayed a modernity of redesigned and reimagined cities. More recently a more knowing late modern urbanism is promoted: for the Brisbane Fair in 1988 an urban promenade was created along the riverside, a place of play and consumption (Figure 4.1). Expo Shanghai

Figure 4.1 Riverside Promenade, Brisbane: the riverside was developed as an attractive public space for the 1988 Expo.

Photo: John Rennie Short.

2010 replaces the urbanism of early modernity by visions of urban sustainability.

World's Fairs, especially during the golden era, reflected anxieties, conflict and contestations in narratives that sought to highlight the wonders of technology, the upward arc of progress, the benefits of industry, the advantages of international trade, and the inevitability of a capitalist modernity. They provided and still provide a dream world of urban fantasies that transform that chaos and disorder of the contemporary city into something coherent, manageable, and managed. Progressive without disturbance, consumption without guilt, and progress without struggle, all wrapped in a built form that intimates a better future.

The Summer Olympics

The modern Olympic Games are the invention of a French bureaucrat Pierre de Coubertin (1863–1937). He drew upon the model of organized sports at US colleges and English public schools that stressed a certain form of masculinity marked by an explicit class bias. The Olympic movement adopted amateurism, essentially as a way to keep out the lower class professional athletes from elite competitive sports.

A certain form of internationalism also influenced Coubertin's worldview. The last third of the nineteenth century saw increased competition between major powers for overseas markets and colonial possessions. This growing interaction in international space initiated a new wave of globalization as new organizations were set up to establish rules for how nations were to interact with each other. The International Postal Convention was signed in 1874, the Hague Convention on rules of war was ratified in 1899 and the International Olympic Committee (IOC) first met in 1894. The first Venice Biennale, which inaugurated the phenomenon (increasingly global and spectacular in recent years) of international contemporary art exhibitions based on the World's Fair model, took place in 1895.

In an earlier work I also examined the Olympic Games, paying particular attention to their international emergence and their intersection with national and urban scales (Short 2004a, pp. 86–108). Here, I focus on the connections between global and local with an emphasis on the Olympics as both a globalizing and urbanizing phenomenon.

The Olympic Games play an important part in the enactment, display and celebration of a global community. From humble beginnings in Athens in 1896 when only 241 athletes from fourteen countries competed with limited press coverage, they have now grown to a truly global spectacle involving most countries in the world (see Figure 4.2, 4.3, Table 4.2). Over the years, the Games have grown in size, scope, and international media coverage. In Beijing in 2008 over 11,000 athletes from 204 countries competed as a global audience watched.

Figure 4.2 The Panathinaiko Stadium: built for the first modern Games in 1896, it was also used in the 2004 Games.

Photo: John Rennie Short.

Figure 4.3 The Olympic Stadium Athens: built for the 1982 European Athletic Championship it was refurbished for the 2004 Olympics including a new roof design by Santiago Calatrava.

Photo: John Rennie Short.

Table 4.2   The Summer Olympic Games

Date	Host city	Participants (women)		Participating countries
1896	Athens	200	(0)	14
1900	Paris	1,205	(19)	26
1904	St Louis	687	(6)	13
1908	London	2,035	(36)	22
1912	Stockholm	2,547	(57)	28
1920	Antwerp	2,668	(77)	29
1924	Paris	3,092	(136)	44
1928	Amsterdam	3014	(290)	46
1932	Los Angeles	1,408	(127)	37
1936	Berlin	4,066	(328)	49
1948	London	4,099	(385)	59
1952	Helsinki	4,925	(518)	69
1956	Melbourne	3,184	(371)	67
1960	Rome	5,346	(610)	83
1964	Tokyo	5,140	(683)	93
1968	Mexico City	5,530	(781)	112
1972	Munich	7,123	(1,058)	121
1976	Montreal	6,028	(1,247)	92
1980	Moscow	5,217	(1,124)	80
1984	Los Angeles	5,330	(1,567)	140
1988	Seoul	8,465	(2,186)	159
1992	Barcelona	9,634	(2,707)	169
1996	Atlanta	10,310	(3,513)	197
2000	Sydney	10,651	(4,069)	199
2004	Athens	10,625	(4,329)	201
2008	Beijing	11,028	(4,746)	204

The Games' growing global reach reflects the shrinking and flattening of the world. An early limiting factor to global participation in the Games was the cost and difficulty of international travel. The high cost of international travel led the Japanese government to offer to subsidize teams' travel to the planned 1940 Olympics in Tokyo. National participation has increased steadily as international travel has become easier and cheaper. The Summer Olympic Games are now one of the most inclusive global events embracing almost every country in the world.

Participation in the Games is now integral to being a member of the international community. It is no accident that countries formerly shunned or marginalized because of their role in international affairs often seek redemption through the Games. The former Axis powers Japan, Italy and (West) Germany all hosted the Games in the post-World War II era as a way back to good standing in the international community. Barcelona hosted the 1992 Games in part to distinguish Spain as a post-Franco democracy and Catalonia as a more autonomous region. And the Chinese government used the hosting of the 2008 Games in Beijing as a way to solidify its role in the new world order (Xu 2006; Schell 2007). The shift from pariah to participatory member can take place by hosting or competing in the Olympic Games.

The Olympic Games are also an important part of political globalization because of the important regulatory role of the IOC. The IOC is one of the first and remains one of the strongest transnational non-government organizations. Coubertin was eager to build an organization independent from the power of individual nation-states and promoted the idea of the Games held in cities in different countries. He envisioned IOC members as ambassadors from the committee to their countries, not as national representatives. Membership was restricted to the wealthy and the well connected. It has remained that way and, although the IOC has grown from fifteen members from thirteen countries in 1894 to 115 members from seventy-eight countries, membership is still biased toward the rich and powerful. The IOC is a self-perpetuating oligarchy immune from democratic accountability. As the Games have grown larger and more financially lucrative, the opportunities for interweaving personal business interests with Olympic concerns have grown dramatically; a number of studies show the strong links between the individual interests of IOC members and Olympic business (Jennings 1996; Jennings and Sambrook 2000; Barney 2002).

The IOC plays a key role in decision making through their control of bidding and hosting the Games. When the IOC decided after the 1988 Games to promote greener, more environmentally friendly Games, cities had to tailor their bids to appear more "green." The IOC now requires all candidates to complete an environmental assessment and actively promotes the concept and practice of a "green" games. Sydney won the right to host the 2000 Games in part because its bid placed greater emphasis on urban reuse and brownfield developments rather than on new build and greenfield developments. The main venue was at the brownfield site of Homebush Bay. Subsequent bids for Summer Games have highlighted more environmentally friendly events and practices. The environment is now the third pillar of the Olympic movement, along with sport and culture. In 2002 the IOC also proposed the encouragement of more compact Games, with venues more concentrated in an urban environment. In the US internal competition to host the 2012 Games, San Francisco lost out because its bid did not conform to the new IOC standard. The IOC's agenda structures the bids of cities, the choice of national Olympic committees when selecting bid cities from their nation and ultimately the character of the resultant Games.

The early Olympics were small affairs with limited media coverage. The first move to the "Games as global spectacle" began with the 1936 Games held in Berlin. The awarding of the Games to Weimar Germany in 1928 signaled the country's return to the international community as the country was banned from the 1920 and 1924 games. Between the awarding of the Games and the opening ceremonies, however, the Nazis gained control of the country. Initially reluctant to host the Games, the Nazi regime eventually saw them as a propaganda opportunity to spread their political message to a wider audience. The 1936 Games inaugurated the torch relay from Olympia. As the torchbearers ran through Austria and Germany, the event turned into a mass demonstration of support for National Socialism. The 1936 Games received multimedia coverage. They were the first Games to use television images, transmitted to twenty-five TV halls in Berlin. Radio transmissions broadcast the Games to forty countries and the Nazis' in-house cinematographer, Leni Riefenstahl, made the film *Olympia* for worldwide distribution. The Nazis' considerable experience in organizing mass events for ideological purposes made the 1936 Games the first truly global spectacle. The opening ceremonies of recent Games have the same sense of mass spectacle.

The Summer Olympics now receive global media attention and television revenue is the biggest single source of income for the local organizing committees and the IOC. Between 1984 and 2008 the IOC generated $10 billion from television coverage rights, comprising almost 50 percent of total revenues. The major television companies now exercise leverage in the staging of the Games. NBC, which paid $3.5 billion for the exclusive rights to show all the Games, summer and winter from 2000 to 2008, requested that the athletic, basketball, gymnastics, and swimming events in Beijing be rescheduled so that they could be shown live on US prime time between 6 and 9 p.m. The IOC agreed to reschedule the gymnastics and swimming events.

The Summer Olympics, in particular, are highly prized by television networks. While very expensive, they can deliver large audiences. Over 3.6 billion individuals watched the 2000 Olympics in Sydney, 3.9 billion watched the Athens Games, and 4.7 billion watched the Beijing Olympics. The proven success of the Summer Games allows the IOC to bundle the less popular Winter Olympics in multi-year deals with the television networks.

The modern Olympics are a commercial event now closely allied to the general process of economic globalization and, in particular, the creation of global audiences of consumers, globally identifiable athletic stars who can be used to sell goods and services in national and international markets, and the increasingly global reach of selected transnational corporations.

Corporate sponsorship began in earnest with the 1984 Los Angeles Games, which was the first Games, since 1896, to be held without government subsidization. The Olympic Partner Program (TOP) was created in 1985. Under TOP, the IOC signs four-year deals with large multinational companies to ensure them global copyright of the use of the Olympic logo in their advertising. Corporate sponsorship continues to expand. In 2001–2004 corporate sponsorship provided close to $1.6 billion, almost 40 percent of total IOC revenues. The corporatization of the Games is explicit in this excerpt from an official Olympic website:

> As an event that commands the focus of the media and the attention of the entire world for two weeks every other year, the Olympic Games are

the most effective international corporate marketing platform in the world, reaching billions of people in over 200 countries and territories throughout the world.

(IOC 2007)

Selected corporations now have a close and symbiotic relationship with the IOC. The Games relies on corporate sponsors, who in turn use the global appeal of the Games as advertising platforms for global economic penetration. While not given any formal voting rights in the IOC, the corporate sponsors wield enormous power. Corporate sponsors pushed to separate the Summer and Winter Games so that they occur in a two-year alternating cycle rather than both occurring in the same year so that corporate advertising expenditures can be more effectively spread out over the fiscal year. To this end, the IOC held Winter Olympics in 1992 and 1994 to inaugurate and synchronize the cycle of an Olympic event every two years. Corporate sponsors also forced the IOC to respond quickly to charges of corruption in the late 1990s. The sponsors did not want to invest in a tarnished logo. They also influence the siting of the Games. There was a persistent corporate emphasis on getting the Games into China in order to gain access to the huge growing market.

Hosting the Games

The Summer Olympics are anchored in individual cities. The Games are named after the host cities and while national contexts are important, the Summer Games are, in essence, an urban spectacular.

In order to forge the link with the idealized Games of the ancient world the first Games of the modern era were sited in Athens. The city did not welcome the Games and a wealthy Greek businessman, George Averoff, reluctantly put up the money to restore the stadium. Soon, however, the early modern Games became an integral part of city boosterism and city image-making.

The second games of the modern era, held in Paris in 1900, were a small and very marginal part of *Exposition Universelle*, a World's Fair. The 1904 Summer Games was simply an added attraction to the St Louis World's Fair and the 1920 Antwerp Games also coincided with a World's Fair. Over the course of the twentieth century, the Olympic Games complemented, then replaced, World's Fairs and Expositions as the new global urban spectacle providing unparalleled opportunities to restructure, reimagine and represent the city.

City elites use the Games in urban boosterism. Roland Renson and Marijke Hollander (1997), for example, show how the elite of Antwerp used the 1920 Games to advance their own economic agenda. More recently, Charles Rutheiser (1996) explores the role of local business elites in promoting the 1996 Games in Atlanta. John and Margaret Gold provide detailed case studies of the relationships between hosting the Games, new urban agendas, and city image-making (Gold and Gold 2007).

Hosting the Games involves the provision of venues and the improvement of urban infrastructure. The trajectory of increasing urban transformation roughly follows a four-stage periodization (Essex and Chalkley 1998). First, 1896 to 1904, was the time of limited urban impacts when the Games consisted of small-scale events with little new urban infrastructure. In the second stage, from 1908 to 1932, the Games grew only slowly in size, still small-scale, but with some new infrastructure. The 68,000-seat White City stadium was the only venue built for the 1908 Games in London, but stood until 1985, an early example of the long-term urban legacies of hosting the Games. The third period, from 1936 to 1976, is marked by the construction of large-scale, purpose-built stadia and major infrastructural improvement. This period opens with the global spectacle of Berlin and the construction of a new Olympic stadium, an elaborate Olympic village and a 20,000-seat swimming arena. It ends with Montreal and an exercise in megastructure modernism involving a purpose-built Olympic Park that housed the new stadium, pool, and velodrome. This era was one of increasing expenditures at a time of limited revenue, before the advent of substantial revenues from major corporate sponsoring and broadcasting rights. As the Games grew in size and complexity, with more athletes needing different types of facilities as well as the growing requirements of press and electronic media, the costs ballooned while revenue stayed low. The Munich Games of 1972, for example, experienced a loss of $687 million (Preuss 2004).

The financial disaster of the Montreal Games, where the net loss, including the cost of all infrastructural investments, is estimated at $1,228 million was a cautionary tale. For the 1984 Games, the US Olympic Committee (USOC) set up a private non-profit corporation, the Los Angeles Organizing Olympic Committee (LAOOC), to make the arrangements so that the city taxpayers were not responsible for the costs. The LAOOC spent around $500 million renovating, rather than rebuilding, Olympic sites. They made $300 million from TV revenue and signed deals with thirty sponsors, who paid between $4 million and $15 million each. Total revenues amounted to $1,123 million, compared to only $467 million in costs. The LAOOC made a profit, corporations achieved global penetration as the Games were broadcast to 156 countries, local businesses made money, and the city became the center of world attention without accruing long-term costs or heavy debt burdens.

The success of the Los Angeles Games inaugurated the fourth era of the Games at a time of rising revenues and increased expenditures involving new and improved venues and major urban infrastructure improvements. The Sydney Games had revenues of $1.2 billion while the Athens Games earned $2.2 billion. The majority of the monies came from television revenues and corporate sponsorship.

Hosting the Games in this fourth era now involves a massive urban transformation including the provision of venues and major infrastructural investment. One stadium sufficed for most of the 1896 Games, but now separate venues are required for swimming, athletics, soccer, tennis, archery, rowing, sailing, kayaking, equestrian events, and many more. As the Games increase in size they make a larger impact on the host cities. The provision of venues is expensive. In order to host the modern Summer Olympic Games in this new era, a city must provide a wide range of sports venues and arenas. Providing all the necessary venues cost Athens $1.2 billion. Beijing spent close to $2 billion on athletic venues, including the $220 million on the 91,000 seat "Bird's Nest" Stadium. Even with massive revenues, venue costs have to be controlled. Many cities now refurbish stadiums as a way to limit costs. The Los Angeles, Seoul, Barcelona, and Athens Games all refurbished their main Olympic stadium. For the 2008 Games, Beijing built twelve new venues, but also created eight temporary venues and refurbished eleven others.

Hosting the Games demands as well as facilitates major urban infrastructural investment. For the 1988 games Seoul expanded its airport, built three new subway lines, and cleaned up the polluted Han River. For the 1992 Games, Barcelona built a new waterfront and upgraded a declining area of the city as well as making numerous improvements throughout the metro area including new roads, a new sewer system, and the creation or improvement of over 200 parks, plazas, and streets. For 2000 Sydney built a new road, linking the airport to the downtown center and constructed the main venue on a previously contaminated inner-city site. Athens spent close to $16 billion in large-scale public investments in water supply, mass transit, and airport connections to get ready for the 2004 Games. Beijing undertook a building frenzy with an estimated $40 billion of Olympic-related buildings and infrastructure including a new expressway and ring roads, miles of rail and subway tracks, and a $2.2 billion new airport that is the largest in the world (Broudehoux 2007; Ren 2008).

The Olympic Games provide both the opportunity and the deadline to implement long-held redevelopment plans. In Greece, for example, the rest of the country has long resisted national investments in Athens. The international showcase of the Olympics provided the necessary spur. The Games start on a specific date with a global audience. The brute reality of such a severe deadline overcomes political resistances, bureaucratic logjams, and administrative inertia.

The Games provide such a huge development potential that urban growth machines mobilize all their powers of influence and persuasion (Andranovitch et al. 2001; Burbank 2001). The idea of hosting the Games has to be sold to urban communities. Backers of the Games routinely induce feelings of community as well as intense nationalism. The huge costs and massive disruption of hosting the Games can pose serious issues for city elites hoping to promote a candidate city. However, the combined economic and political weight of the "urban growth machine" pushes past opposition. Burbank et al. (2000) show that citizen opposition to Olympic-related economic growth in a selection of cities was piecemeal rather than sustained and effective. There is an asymmetry in power and resources between urban elites and local opposition. Resistance may not be futile, but it is rarely successful, as those powerful forces advocating hosting the Games hijack the narrative. And underlying the limited opposition to this narrative is the fact that the Games retain a strong positive image. In a time of fraying civic connections, bidding for and hosting the Games holds both the promise and reality of creating a sense of solidarity, a feeling of communal ownership of the event and a collective goodwill during and immediately after the Games. The Games tap into large reservoirs of civic pride and deep feelings of urban community. Peter Newman (2007) shows the wide popular support for the London bid. The bid, and the prospect of hosting the Games, provided a celebratory framework that generated substantial popular as well as business support.

Because the Summer Games are now in a gargantuan mode with huge revenues and vast expenditures the obvious question to ask is: are the Games a net cost or a net benefit to the host city? An accurate answer has proven elusive. Inclusive cost–benefit analyses are severely hampered by lack of proper accounting methods, technical issues such as accurately estimating exchange rates of foreign currencies and such basic issues as the lack of available data. Costing the Olympic Games is a fiscal mystery made more complex by the partisan nature of much of the analyses. Organizations seeking to promote, justify or attract the Games generate the vast majority of cost–benefit analyses. While many studies highlight the positive benefits of the Games (largely funded by organizations and groups seeking to justify the Games), fewer studies examine the costs of the Games and the redistributional consequences (but see Lenskyj 2000, 2002; Waitt 2008). The Games involve massive investments that crowd out other forms of public investment, such as spending on education and social welfare that may serve better the long-term needs of ordinary citizens. Other costs include the dislocation to the city during the construction period and the possibility of increased house prices and rent levels. There are also social and environmental costs that are rarely factored into the standard cost–benefit or economic modeling approaches. A recent report estimates that the staging of the last twenty Olympic Games displaced twenty million people (Center on Housing Rights and Evictions 2007). The report also highlights the displacement of almost three-quarters of a million people for the Seoul Games; 30,000, predominantly African Americans, for the Atlanta Games; and approximately 1.25 million displaced for the Beijing Games.

Cost–benefit analysis of hosting the Games remains at a rudimentary stage, with few accurate or comprehensive studies and little comparative data. There is often massive public investment in the Olympic Games that often goes unrecorded in most cost–benefit analysis. The existing analyses suggest that the distribution of costs and benefits is regressive with most of the costs borne locally, especially by the more marginal urban residents displaced to make way for the Games, while most of the benefits accrue to local elites and a global media market. The measurement of costs and benefits and their social distribution deserve much more careful analysis because there is systemic overestimation of benefits and underestimation of costs. While some of the cost overruns may be due to unforeseen events such as changing IOC requirements, their persistence suggests intentional low-balling of bids and the generation of cost–benefit analyses more sensitive to benefits than costs (Matheson 2002; Bergen 2007).

What are the urban economic impacts of hosting the Games? The Games have a wide economic impact on jobs and economic activity from the pre-Games phase of construction, through the intense activity of the two-week Games and into the legacy of the post-Games. City economies experience direct effects of new money spent by outside visitors, indirect effects of the injected money, and induced effects of increased income. A survey found that hosting the 1996 Atlanta Games boosted employment by as much as 17 percent in selected counties in Georgia (Hotchkiss et al. 2003). Evangelia Kasimati (2003) reviews the economic impact literature and finds that the Games result in economic growth, increased tourism, and additional employment. One report estimates that the net economic impact on the host city averages from $4 billion to $5 billion (McKay and Plumb 2001). Holger Preuss (2004) summarizes a vast amount of economic data to reach the conclusion that, since 1972, all Summer Games, excluding investments, made a surplus. The net balance of Olympic Games' revenues and expenditures was, on average, a surplus of $300 million.

Preuss (2004) suggests an economic legacy of a small financial surplus, structural improvements to the city including new and improved sports facilities, and enhanced international image in the wake of a successful hosting of the Games, which results in increased tourism and foreign investment. However, each of Preuss' conclusions is subject to some scrutiny. The financial surplus conclusion, for example, often refers to the surplus in the revenues that the local organizing Committee generates over what they expend in operating and infrastructure investment. It often ignores or downplays the massive public investment in the hosting of the Games. The Atlanta Games, for example, reported a small surplus yet this ignores the approximately $2 billion spent by public authorities including $996 million in federal government investment, $226 million in state funds, and $857 million in local funds. The costs are borne by public authorities while the revenues are privatized by private or non-profit organizations that are neither democratically elected nor publicly accountable.

Even the legacy of substantial sporting facilities can be a mixed blessing. The modern Olympics require massive investment, often concentrated in a large central site. This represents a huge space to maintain and use effectively and efficiently after the Games are ended. Built to house hundreds of thousands, the main Olympics sites in Montreal, Sydney, and Athens, for example, are rarely used to full capacity and are a net drain on resources. Glen Searle (2002) shows that the two main Olympic stadia in post-Games Sydney experienced major revenue shortfalls caused by competition from other smaller stadiums and a lack of major sporting events. The main site of the Athens Olympics costs close to $500,000 per year in just the basic maintenance of heating, cooling, lighting, and security. It will take thirty years to pay off the cost of building the Bird's Nest; meanwhile it costs $19 million a year just in maintenance and debt repayments. Authorities are seeking to sell the naming rights to offset costs. The massively expensive sporting facilities necessary to host the Games are often underutilized and expensive to maintain after the Games are over.

Even tourist benefits vary by location. Long haul destinations such as Sydney are differentially impacted than short haul destinations such as Barcelona, which saw a more sustained increase in post-Olympic visitors and tourists. Hosting the Games also produces displacement effects with some tourists turned away by the crowds and traffic in host cities. The relatively low ticket sales in Atlanta, Sydney, and Athens suggests that the Olympics are becoming more of a global media event, than a real-time, real-place event attended physically by visitors.

Big games, big cities

In the immediate post-World War II era, US cities, because they had the resources and had avoided war-related damage, dominated the pool of candidate cities. Of the seven candidate cities for the 1952 Olympics, five were US cities: Chicago, Detroit, Los Angeles, Minneapolis, and Philadelphia. Detroit was a candidate six times from 1952 to 1972.

As the costs of the Games escalated and revenues stagnated, the number of candidate cities dropped off. There were only four candidate cities for the 1964, 1968 and 1972 Games, three for 1976, two for 1980 and only one for 1984. The financial success of Los Angeles inaugurated a new era of intense city competition to host the Olympic Games. Table 4.3 lists the bid cities since 1976. Since the 1984 Games, and especially since 1988, when the fiscally positive experience of the 1984 Games was factored into Olympic bids, the number of candidate cities has increased and now includes a wider range of cities outside Western Europe and North America. However, although the number of candidate cities has increased over the past twenty years, not all cities make the final rounds of IOC deliberations. Table 4.3 also notes the cities eliminated from the final vote. As the Games become more elaborate and more expensive with a greater reliance on already-existing substantial urban and sporting infrastructure, many cities in small and/or poor countries find it difficult to make it past the early rounds. Witness the early rejection of Havana for the 2008 and 2012 Games, and the fate of Baku compared to Chicago or Tokyo in the bidding process for the 2016 Games. Financial ability to host the Games and commercial "spin-off"(i.e. the ability to use the Games to penetrate new markets) are principal considerations for the IOC. Beijing was almost certain to win, given the willingness of the Chinese government to fund the Games and the huge commercial potential of the Chinese market to corporate sponsors.

Table 4.3 Candidate and host cities, 1976–2016

Year	Host city	Bid city
2016	Rio de Janerio	Chicago, US Madrid, Spain Rio de Janeiro, Brazil Tokyo, Japan Baku, Azerbaijan* Doha, Qatar* Prague, Czech Republic*
2012	London	Paris, France Madrid, Spain New York, US Moscow, Russia Leipzig, Germany Rio de Janeiro, Brazil Istanbul, Turkey* Havana, Cuba*
2008	Beijing	Toronto, Canada Paris, France Istanbul, Turkey Osaka, Japan Bangkok, Thailand* Cairo, Egypt* Havana, Cuba* Kuala Lumpur, Malaysia* Seville, Spain*
2004	Athens	Rome, Italy Cape Town, South Africa Stockholm, Sweden Buenos Aires, Argentina Istanbul, Turkey* Lille, France* Rio de Janeiro, Brazil* St Petersburg, Russia* San Juan, Puerto Rico* Seville, Spain*
2000	Sydney	Beijing, China Manchester, UK Berlin, Germany Istanbul, Turkey
1996	Atlanta	Athens, Greece Toronto, Canada Melbourne, Australia Manchester, UK Belgrade, Yugoslavia
1992	Barcelona	Paris, France Belgrade, Yugoslavia Brisbane, Australia Birmingham, UK Amsterdam, Netherlands
1988	Seoul	Nagoya, Japan
1984	Los Angeles	
1980	Moscow	Los Angeles
1976	Montreal	Moscow Los Angeles

* Indicates cities that did not make it to the final round.

Candidate cities now include global cities at the apex of the global urban hierarchy (Shoval 2002). Major cities such as New York, London, and Paris now compete to host the Olympics and National Olympic Committees now increasingly promote the larger cities as their candidate cities.

Cities become candidates after winning approval from their National Olympic Committees. Each country has only one candidate city. The larger, richer countries have an internal competition. The competition for the US candidacy for the 2012 Games began in 1997 with eight cities—Cincinnati, Dallas, Houston, Los Angeles, New York, San Francisco, Tampa, and Washington-Baltimore. The USOC dropped Cincinnati, Dallas, and Los Angeles from consideration. In February 2002 only four cities made the cut: Houston, New York, San Francisco, and Washington-Baltimore. In November the decision came down to New York and San Francisco. New York was selected. The city's global recognition was an important factor. The USOC eliminated Houston from the final consideration because of "poor international recognition." The National Olympic Committee of Great Britain and Northern Ireland proposed Manchester for the 1996 and 2000 Games but was only successful when it put forward London as its candidate city for the 2012 Games.

Because of the heavy infrastructural requirements of hosting such a large international event it is only the larger and richer cities that are serious candidates to host the Summer Olympics. Small cities and cities from developing countries tend to be cut early from the bidding process. And while the refurbishment rather than the new building of sporting venues does limit costs, it also has the effect of limiting successful and viable bids to large, relatively wealthy cities that already have a substantial sports infrastructure. The possibility of cities from poor developing countries with limited existing facilities hosting the Games is rapidly diminishing in this new era.

Let us consider the competitive process to host the 2012 Games. It began with a formal request in May 2003 by the IOC for National Olympic Committees to submit a candidate city. All applicant cities had to submit a fee of $150,000 by August 2003. The nine nominated cities—Havana, Istanbul, Leipzig, London, Madrid, Moscow, New York, Paris, and Rio de Janeiro— then had to reply to an IOC questionnaire asking about the long-term impact and legacy of hosting the Games, the Games' place in the long-term planning of the city, the level of public and government support, the existing and planned sporting infrastructure, and the environmental impact of the Games. Based on the responses, the IOC drew up a shortlist in May 2004 that included London, Madrid, Moscow, New York, and Paris. These cities then had to submit a more extensive three-volume bid book by November 2004. Each of the shortlisted received a three-day visit in February–March 2005. The IOC took a final vote in July 2005 in Singapore. Making a serious bid is an expensive proposition. The successful London bid cost $25 million.

The global city imaginary

Across the world, city elites are promoting a global city imaginary; a vision of a self-consciously "modern" or "global" city—the terms are used interchangeably—revealing the connection between modernity and globalization, replete with images of busy international airports, foreign tourists, inward investment, a cosmopolitan atmosphere, creative industries, cultural economies, and an overwhelmingly positive image shared around the world. This image is used to promote and justify all manner of policies and programs. Elites and urban political regimes call on this imagery to generate public support for projects, to justify tax incentives and business-friendly policies in an era of neoliberal economics and sometimes to support multiculturalism and cosmopolitanism. Bidding and hosting the Games can become an important element of this discourse. The Games provide an opportunity for massive urban renewal, urban restructuring, major environmental remediation, dramatic infrastructural improvements, and the creation of a positive image across the globe. Hosting the Games is winning the gold medal of global intercity competition.

There are paradoxes in the employment of this imaginary. Local residents, for example, are often encouraged to support Olympic bids with the rhetoric of "putting the city on the world map," a campaign of making and reaffirming its global city status. However, the example of Houston's failed bid to host the 2012 Games is a salutary reminder of a city's limitations. The USOC rejected the bid because the city did not have a high enough international profile. I will refer to this as the "Houston paradox." In other words, the very cities that are desperate to host the Games, because they are in the wannabe global city category, are the ones most likely to fail.

However, even unsuccessful bids have an important role in a reimagining of the city. The Baltimore-Washington bid for the 2012 Games failed in the early US rounds. A central theme of the bid was the remediation of the Anacostia River. Popular support was mobilized, and although it did decline it did not disappear after the bid's rejection. Even unsuccessful bids can generate new ways of imagining a city.

A successful bid gives a city increasing odds of hosting other events. A city that hosts the Games increases its stock and value of athletic venues, improves infrastructure and creates new political alliances and positive imagery that it can use in subsequent competitions. The city now has traction in its bid to host other mega-events. That Seoul and Korea hosted the 1988 Games allowed it to successfully compete as one of the sites for the 2002 World Cup. With new facilities, a proven track record of organization, growth coalitions already in place, new infrastructure and name recognition, an Olympic host city can successfully compete for other mega-events. Hosting the Games puts a city on a new trajectory of global competitiveness as Olympics success breeds even more success. In subsequent years we may see a small group of elite cities emerge as the exclusive top tier hosts of global mega-events.

Hosting the Games provides an opportunity for city reimaginings as well as remakings. In the context of a global media spotlight, the spectacular staging of the Games becomes the setting for a dramaturgical remaking and representation of the city. Three narratives dominate: the creation of a green city, the making of a modern city, and the construction of a global city.

Growing recognition of global warming and the emergence of an urban environmental movement now links with a city's "green-ness" as a sign of urban competitiveness (Benton-Short and Short 2008). Not to be green is now similar to being an industrial city, a sign of the past compared to the future, the old in contrast to the new. A green city is closer to the unfolding future than to the disappearing past. To be considered a green city, despite all the ambiguities and difficulties in actualizing such a term, is to be seen as globally competitive. Beijing's continuing air pollution, despite all the massive efforts to minimize it for the Games, literally and metaphorically cast a pall over the city and undermined its claim to be a truly global city.

Beginning with Sydney there has been a deliberate greening of the Games. This Olympic trope now coincides with a more general global imaginary of a green city. The temporary and permanent greening of the city is a possibility since the IOC now promotes "green" games. The Sydney Games were held on a former abandoned waste site, Homebush Bay, and the Games were part of a wider urban remediation and greening project. Solar panels provided energy to the Olympic Village. To reduce air pollution the organizers of the Beijing Games closed almost 200 factories and 680 mines and undertook an extensive reforestation program. Beijing also upgraded its public transportation and replaced its old 18,000-bus fleet with more fuel-efficient and less polluting vehicles. The city government spent $13 billion on environmental clean-up. For the 2012 Games, London set aside almost $2 billion out of total costs estimated at $4.7 billion for redeveloping the run-down and largely abandoned Lower Lea Valley.

The Olympics also provide an opportunity to make a modern city. Athens, for example, upgraded its infrastructure by installing a new transportation network, including motorways, a tramway, and a metro system. The construction of a pedestrian walkway pulled together the ancient archaeological sites of the city center in one sweeping promenade. Elias Beriatos and Aspa Gospodini (2004) describe the creation of a "glocalized" landscape that involves the promotion of built heritage as well as innovative design. Before the Olympics, Athens lacked coherent development plans; it was a disorganized capital popularly represented as a city on the verge of suffocating in its own chaotic growth. The renovation of the seafront, the organization of the historic center, the cooperation between public and private agencies and the new use of metropolitan-wide government all became possible. Hosting the Olympics provided the opportunity for city-wide, coherent planning to create a modern city. Modernity can come at a cost. The Chinese government used the 2008 Games as an opportunity to modernize Beijing. One plan involved the destruction of the old, high-density neighborhoods of small alleyways in the central part of the city, seen by officials as a remnant of a premodern past. Almost 20 square kilometers were destroyed and almost 580,000 people were displaced in this one program (Fan 2008a). Ai Wei Wei, one of the designers of the Bird's Nest, later became a vocal critic of the old neighborhood destruction in Beijing. His monumental sculpture made from the doors of the destroyed houses of the Beijing alleyways (*hutong*) was featured at the 2007 Documenta, and it functions as an arresting counterpart to the Bird's Nest.

The creation of a more global city is also one of the main goals and consequences of hosting the Games. In some cases the goal is explicit, as in the case of the South Korean government's plan to use the 1988 Olympics as a way to globalize Seoul and open the country up to the wider world. The Games allow a city to showcase itself to a global audience and become more globally connected. For two weeks the city is constantly mentioned and represented in the world's media. New or greatly expanded international airports are standard requirements, as are new road and mass transit linkages from the airport to the rest of the city. Hosting the Games allows the city to achieve global recognition with the possibility of increased tourism and investment.

The relationship between the Summer Games, modernity, and globalization is both simple and complex; simple in the sense that the increased participation in and consumption of the Games embodies a more globalized world. Most countries compete in the Games, and most television audiences watch the Games. The easy conclusion is that the Games are now a truly international event watched around the world. However, when we look at different aspects of globalization in more detail, the picture is more complex. In terms of political globalization, the Games do not transcend nationalism but reinforce it. The national anthems, the national uniforms, and all the national iconography all define the Games as an international arena for the celebration of nationalism, not its transcendence. In terms of cultural globalization, the Games do provide a common set of images, a shared cultural experience. But when we look in detail at television coverage we see how different audiences watch different Games. The opening and closing ceremonies are shared but much of the rest of the television consumption of the Games is tailored for separate national and regional audiences. There is one form of globalization, however, where the relationship is more direct. The Games have provided a platform of economic globalization as selected multinational companies get their brand name connected with an event that penetrates most markets in the world. John Horne and Wolfram Manzenreiter (2006) write of a sport–media– business alliance. The Games are an important event for this alliance because they offer unparalleled opportunity for selected multinational companies to widen their target audiences and extend their global market share.

In the urbanizing of the Games we see a general trend for increasing competition to host the Games and increasing impacts on the host city. The competition to host the Games has increased because there is a widespread belief, strongly held by many city business elites, that hosting the Games can be of positive benefit by providing revenue, a legacy of infrastructure, and a new global recognition to cities. Hosting the Summer Games now provides an incentive and opportunity for city elites to restructure their cities in an increasingly competitive environment. They allow urban makeovers designed to create a modern city defined in terms of a firmer insertion into international circuits of capital flows and the construction of modern urban forms. The changes often come with increased costs for city residents unless there is a specific commitment to redistributional outcomes.

In a globally competitive world, hosting the Summer Olympics allows a rare opportunity for a major restructuring of the built form, a major economic boost, an increased global connectivity, the creation of an urban modernity and a profound change in the global image of a city. In the immediate aftermath of a reputedly $40 billion Beijing Games, the stakes are raised again. The lavish spending, the grand spectaculars of the opening and closing ceremonies all ratchet up the expectations for subsequent Summer Olympics. Yet the competition continues. Bidding for and hosting the Summer Olympic Games are active moments in the global city imaginary, that constellation of ideas and practices associated with the pursuit of global city status and the creation of a truly modern city.

CASE STUDY 4.1 **Centennial Park, Atlanta: a legacy of the Games**

"It's Atlanta." The bold face type dominated the front page of the *Atlanta Journal* on the morning of September 18, 1990 after the IOC announced that the city would host the 1996 Centennial Olympic Games. The next line in the headline went on, "City explodes in thrill of victory." Further down was the sense of relief "We finally won something."

The Atlanta Committee for the Olympic Games (ACOG) was the privately funded not-for-profit organization responsible for organizing the 1996 Centennial Games.

ACOG invested $500 million in venue construction out of a total budget of $1.7 billion. There was also a 1996 Olympic Arts Festival that involved 200 theater, dance, classical music, and jazz performances, twenty-five exhibitions and seventeen public art works. Entertainers included the Russian National Orchestra and the Miami City Ballet. A new play by Sam Shepard was performed.

"Georgia will continue to benefit long after the Olympic Games end, as businesses are drawn to our state by Atlanta's growing reputation as an international business center," noted the Olympic Games Staff Handbook on page 16. In reality, the permanent legacy of public spaces was relatively slight in Atlanta compared to Barcelona. Many of the venues reverted to private spaces. The 85,000-seat Olympic Stadium became the 52,000-seat home of the Atlanta Braves baseball franchise. The Atlanta University Center, a complex of six old

Figure 4.4 Centennial Park, Atlanta.
Photo: John Rennie Short.

historically black colleges obtained the 6000-seat gymnasium and two hockey fields. The Aquatic Center and athlete housing became the property respectively of Georgia Tech and Georgia State University.

A rare exception was Centennial Olympic Park, a new 21 acre public space in the downtown. The Park, with its five-ring fountain was handed over to the state of Georgia after the games and has evolved into a multipurpose-use event space that was used as a catalyst for further downtown investment and development, and the entertainment branding of downtown Atlanta including the World of Coke and the Georgia Aquarium. It hosts hundreds of events and receives over a million visitors a year. It is one of the very few public legacies of a Games dominated by corporate interests.

CASE STUDY 4.2 **Mega-events, spectacles, and the city: the case of Barcelona**

The hosting of three urban spectaculars, two World's Fairs in 1888 and 1929 and the Summer Olympic Games in 1992, impacts Barcelona's urban form. And these events also shape debates on urban modernity.

The 1888 World's Fair, known as the Universal Exhibition, took place at a time when Barcelona was growing as an industrial city. There were so many factories it was called the Lancashire of Spain. There was also rising class conflict. The first general strike took place in 1855. The rich left the old city. The Eixample (Extension) was built in the 1860s on a grid system, the rational order a direct response to the convoluted street pattern of the Gothic city. New urban forms represented the modern resistance to the weight of the past as well as a way to restructure a tightly compacted, socially conflicted city. There was also a cultural renaissance in the region, the Renaixenca, Catalan Rebirth, which emerged from centuries of suffocating Castilian dominance. The first daily newspaper in Catalan was published in 1879.

It is in this context of economic and political change and cultural renaissance that the 1888 Fair took place. Proponents argued that it would show the city as European, code for modern, rather than Spanish, a codeword for backwardness.

The Exhibition if not always causing, at least facilitated urban transformations and new architectural practices. There were the tall structures; in this case the monument to Christopher Columbus that still stands at the bottom of the Ramblas. There were the consolidations of route ways through the city. The main site was built on the site of a fortress used by the Spanish army to control the city in another symbolic distancing of the Catalan capital from Castilian Spain. Large buildings were laid out in a horseshoe design in a park-like setting with a triumphal arch framing the entrance. A number of the buildings, especially those by Montaner, were designed in an early form of Art Nouveau, enthusiastically embraced and referred to as "*modernisme*" in Barcelona. Called at short notice, the Fair was not a great international success. There were few foreign visitors and many exhibitors were waiting for the 1889 Expo in Paris. However, two-and-a-half-million Catalans visited the Fair and witnessed its embrace of modernity, the trust in technology framed by Art Nouveau aesthetics, and the politics of Catalan nationalism. Barcelona became a significant center for *modernisme*: the architecture of Antoni Gaudi, for example, grew and blossomed in the city.

The second event was the 1929 International Exhibition. A number of changes were prompted and facilitated by the event, including the construction of Plaza España and the first underground railway. The main site was at Montjuic close to the 1888 site. There was some urban renewal as the area around Plaza España was cleared of low-income housing. Modernism was active and on display. The German Pavilion, for example, was an elegant building designed by Mies van der Rohe. The event again showcased the city to the world, prompted urban make-overs, and reinforced a modernist urban aesthetic.

In 1992 the city hosted the Summer Olympics. The main site was also at Montjuic. Signature architects from around the world were employed. But unlike Atlanta in 1996, the city was transformed with an eye to upgrading public space and improving the lives of the citizens. A new long-needed ring road was constructed. Small neighborhood parks and larger public open spaces were upgraded and improved. Barcelona undertook an ambitious urban renewal program that involved turning the old docks and railway yards into a revitalized public space that is still used by local people long after the Games have ended. The Olympic Village was built in Poblenou, an area dominated by abandoned

Figure 4.5 Barcelona: part of the old port was opened up for the 1992 Olympics extending the walkway of the Ramblas from Plaza Catalunya to the Mediterranean.

Photo: John Rennie Short.

industrial sites. The makeover involved the demolition of old buildings as well as the construction of new retail sites, hotels, and office space. Since 1992 the area has attracted more middle-income households and has become one of the new growth areas of the city (Lambiri 2005).

Barcelona successfully used the Games to reposition itself as a major tourist destination, almost doubling the number of foreign visitors two years after the Games compared to two years prior to the Games (Brunet 2005). The Games transformed the city and improved the image of their city. It is now considered one of those "cool" cities, its architecture and public space drawing people from around the world. The Barcelona Games have become a yardstick for Olympics urban makeovers that improve the city for most of its residents.

5 *Flânerie* and the globalizing city

There is a pressing need to realize a more embodied discourse that incorporates the considerable physical and emotional dimensions of the human body's interaction with global, urbanized space. In this chapter I review the history and current revival of *flânerie*, assessing it as a lens for understanding and representing cities undergoing globalization.

Historically, the *flâneur* is considered an iconic figure who processed the newly modern city of the nineteenth-century industrializing West (particularly Paris) in subjective terms by randomly wandering through it, absorbing its shocking novelties, and translating his experiences into art that was as radically modern as the fleeting moments it attempted to capture. Impressionism is considered to be a quintessential product of *flânerie* as well as much of early avant-garde cinema devoted to kaleidoscopic urban life. As urban life became more mundane, as subjective expressivity became suspect, as cosmopolitanism seemed a naïve, utopian Eurocentricism, *flânerie* faded away as yet another passé modernist trope.

This was the intellectual arc from the 1860s to around the 1980s in Western scholarship. For the humanities, *flânerie* was largely a nineteenth-century practice whereby artists derived inspiration from their direct, subjective experiences of the newly industrialized, consumerist city. For the social sciences, *flânerie* was largely a twentieth-century practice whereby urbanists derived knowledge from an ambulatory reading of the city as text. The former was highly emotional and multisensual, involving random strolling. The latter was affective yet more dispassionate and primarily visual, involving a strategic, ethnographic mobility. Both traditions were conceived in terms of traversing Western cities on foot and suffered as urban spaces became less conducive to such locomotion. The seeming deathblow for *flânerie* as a means toward urban understanding seemed to occur when cities' spaces became so transparent through planning and commodification as no longer to afford sufficient spontaneity for the practice's humanistic stripe nor to demand the intense scrutiny of its sociological stripe. Simultaneously, emerging postmodern perspectives questioned the validity of urban analysis based upon affective, embodied experiences and especially upon Western cities alone. Notions that developed in tandem with the rise of the metropolitan West, such as *modernism* and *cosmopolitanism*, were dismissed as too narrow to have meaning for the rest of the world and inappropriate as urban universals. *Flânerie* was similarly tainted. However, the connections between the Third Urban Revolution and the Late Modern Wave of Globalization have resuscitated, re-energized, and reshaped *flânerie*'s contemporary relevance. Today artists and social commentators of both genders have realized the need for a practice that can embody a sensuous, direct, immediate, and emotional apprehension of a modernizing globalizing city.

The original *flâneur*

The French verb *flâner* is the root of *flâneur* and *flânerie*; it means "to stroll, to loiter, to dawdle," and initially reflected negative assumptions about the practice as mere wasteful idling (Ferguson 1994). The original *flâneur* wandered leisurely through the passages of rapidly growing, industrializing cities of the nineteenth century. The unhurried gait was deliberately deceptive, however: it was actually a cover for the *flâneur*'s true mission to tap eagerly into urban energies, to absorb readily unforeseen novelties, and to engage playfully (yet ruefully) the mingling of old and new typical of modern cities. The consummate *flâneur* created art from these practices. The *flâneur*'s original proving ground was rapidly urbanizing Paris, designated by Walter Benjamin as the "capital of the nineteenth century." Traversal of crowded Paris, while absorbing its affective intensities for aesthetic translation, constituted the practice of *flânerie*. The predominant use of the noun's masculine gender suggests the absence of a *flâneuse*, although recent scholarship asserts the *flâneuse* as a distinct, if limited, reality on the streets of nineteenth-century Paris (D'Souza and McDonough 2006).

Baudelaire's 1863 essay, "The Painter of Modern Life," elevates the *flâneur* from a marginal, indolent ambler—"mere *flâneur*"—to a substantive artist-*flâneur*—"perfect *flâneur*." For Baudelaire, the artist-*flâneur* was a connoisseur of the details of everyday metropolitan existence as well as a producer of art from the direct observation, gauging, and interpretation of urban dwellers. By doing so, he engaged in the "heroism of modern life" (Baudelaire 1965, pp. 118–119). For Baudelaire, the heroic response to the modern condition involved willingly weathering its shocks for its gift of kaleidoscopic consciousness, the key to an urban-based creative production, as it could reproduce "the multiplicity of life and flickering grace of all elements of life"(Baudelaire 1970, p. 9). Throughout the essay, Baudelaire maintains that the true painter of modern life suppresses any lingering fears of drowning in urban banality and surrenders to a joyous frenzy of inhaling as much of the redolent street blend as possible before it disperses. Making art out of modern Paris required that the artist-*flâneur* conjure lasting beauty from ephemerality's fleeting, chaotic details. Baudelaire did not regret modernity's constant passing into memory. Indeed it is clear from his writings that fleeting experiences amidst the urban evoke immediate pleasure as well as elegiac melancholy.

Baudelaire enlists phantasmagoria and kaleidoscopes to evoke the sensual bombardment of urban experience. Appearing around the turn of the nineteenth century as ingenious mechanical devices of optical display, "phantasmagoria" and "kaleidoscope" possessed undeniable ocular resonance as referents. Baudelaire, the latent Romantic, employed the haunted apparitional as metaphor of Paris' spellbinding power. Yet Baudelaire, the nascent Modernist, redefined phantasmagoria's and kaleidoscope's special effects as less horror-scopic in his attempt to capture the sensory strength of modernizing Paris' new, spontaneous, contingent, ongoing street extravaganzas. Baudelaire's writing reflects a rich phenomenology of experiencing the urban everyday with wonder. He recommended the practice of this impulsive, immersive creative act, which he famously likened to a "bath of multitude" and a "debauch in vitality" (1970, p. 20). He channeled the gustatory, olfactory, and tactile in particular, producing an urban art of a profoundly multisensate, lived experience.

Walter Benjamin

Walter Benjamin is regarded as *flânerie*'s pre-eminent theoretician with considerable justification since he grappled mightily from 1929 to the end of his life with what *flânerie* could and could not mobilize in response to urban modernity's onrush, alternately celebrating and decrying the potential of the *flâneur*'s abilities. Out of what is unmistakably a very keen identification with the circumstances of Baudelaire's art and life, Benjamin appropriated the *flâneur*'s experience in what has been aptly described by Michael Jennings as a "powerful projective act" (Benjamin and Jennings, 2006, p. 16).

Benjamin is one among a series of critical thinkers about the city whose influence has grown in the last few decades. A large body of Benjamin's work was translated into English for the first time throughout the 1970s and 1980s, and a larger body of Benjamin scholarship followed soon after. Then in the 1990s, *Das Passagenwerk* or *The Arcades Project* (Benjamin 1999, henceforth AP), Benjamin's ever in-progress compendium of notes and fragments about nineteenth-century Paris, was published in their relative entirety for the first time in German and not long after in English. In the 2000s, Benjamin scholarship exploded.

Throughout AP, Benjamin often vicariously relives Baudelaire's explorations of nineteenth-century Paris, and when he does so, he also celebrates nineteenth-century Parisian *flânerie*'s embodied, sensuous, and subjective side—its Baudelairean incarnation. More often, however, Benjamin mourns the *flâneur*'s demise as Hausmannization managed and therefore restricted the *flâneur*'s creative wandering through the streets of Paris, thus sealing his fate as a mere consumer, rather than producer, of culture. For the most part, Benjamin was pessimistic about the aesthetic yield of *flânerie* for contemporary artists as consumerist contrivance increasingly managed urban life. Benjamin's lament won the scholarly day, and the prevailing view for most of the late twentieth century was that the *flâneur*, Baudelaire-style, incrementally lost the ability to make art out of the city. Instead, according to Benjamin, he walked around gaping, merely window-shopping, on the brink of extinction.

Benjamin's intellectual heirs have extended his notion of the managed *flâneur*'s uninspired perambulation as superseding the Baudelairean *flâneur*'s sensuous embodiment of the modernizing city: they now cite contemporary cities' increasingly prescribed routes, commodified pathways, and "heritage walks" as having drained the immediacy and experiential from the urban environment by replacing the former realities of the streets with simulacra. In the process, true *flânerie* is managed to the point of near nonexistence. This assessment relates primarily to Western and colonial cities.

However, amidst AP's general gloom about the *flâneur*'s tragic fate, Benjamin also outlines a possible redemption of the Baudelairean *flâneur*'s embodied, sensuous, spontaneous side so long as it manifests when the *flâneur* is in the midst of penetrating consumerist spectacle rather than just uncritically immersing in it. The nature of the redemption of the Baudelairean artist-*flâneur* for Benjamin's own purposes is exemplified by his appropriation of the phantasmagoria concept. Both Baudelaire and Benjamin agree the phantasmagoria captures the urban spectacle as a dizzying yet spellbinding overlap of architectural demolition and construction, of past and present. Both agree that phantasmagoria's shock can be pleasurable. Yet for Benjamin, this shock must lead to a dialectical emancipation that will rescue the artist-*flâneur* from death-by-commodity. For Benjamin, the goal is not to revel in the grasp of phantasmagoria but to break through it into a calmer, more reflective state of consciousness, one that can resist urban modernity's consumerist capitalist spell (Wolin 1982). Throughout the last quarter of the twentieth century, the concept of Benjamin's reflective *flânerie*'s resistance to urban modernity prevailed more than Baudelaire's reflexive *flânerie*'s encounter with urban modernity.

Benjamin's work is so encyclopedic, so complicated, so unfinished, and so often contradictory that it just endlessly spins itself as source material for a variety of interpretative goals. We do not deny Benjamin's important bequest to urban aesthetics, especially in terms of their ethnographic dimension; to useful notions of the everyday; to the foregrounding of *flânerie* by solidifying its connection to urban modernity; and to his inspired riffs on allegories of the urban. Yet, Benjamin's extracting surreal potential from the Parisian past to accord with his view of urban modernism is of limited value to our notions of the urban experience, past or present. In "Walter Benjamin's Myth of the *Flâneur*" Martina Lauster (2007) makes a convincing case for Benjamin's highly idiosyncratic and 1930s-oriented perspective as blind to a great deal of nineteenth-century Paris' "journalistic" everyday when it comes to *flânerie*. She contends that this blindness has been adopted by "Benjamin-inspired cultural theory" to the detriment of more empirically founded approaches to the city. Although Lauster's focus is primarily upon nineteenth-century interpretations of the city, her insights imply that Benjamin is also problematic to the intensely present-oriented experiences of twenty-first-century urban places. Lauster's work is a prime example of recent scholarship more prone to questioning whether Benjamin's predominant theorization of *flânerie* truly is germane to today's globalizing cities beyond the West and whether his considerable insights into nineteenth-century Paris are necessarily relevant (Sarlo 2001; Chandler and Gilmartin 2005). This scholarship is often skeptical of Benjamin's wholesale application to the present. Persistent questions refer to Benjamin's frequent dismissal of the positive aspects of the growth of mass culture and whether or not his *flâneur* has so utterly morphed under the conditions of global modernity so as to be unrecognizable (Goebel 2001; Gilloch 2002; Richter 2002; Gumbrecht and Marrinan 2003; Spiridon 2005).

Resuscitating pre-Benjamin flâneurs

Parallel to a more searching critique of Benjamin is a flâneuristic resuscitation of pre-Benjamin urban commentators. Friedrich Engels's *Condition of The English Working Class in 1844*, for example, has been retrieved by the new scholarly paradigm that establishes aesthetic reflexivity as a guiding principle for the intellectual cum sensual navigation of globalizing cities. In "'More than Abstract Knowledge': Friedrich Engels in Industrial Manchester," Aruna Krishnamurthy (2000) links Engels's ambulatory spatial practices with lived experiences of Mancunian working class culture. In doing so, she confers ethnographic, even empathic, reflexivity on Engels's walks around Manchester. Engels's descriptions of Manchester's new urban forms throughout *The Condition of the Working Class in England* are rife with aesthetic power, revealing a considerable ability to draw upon artistic means, his own as well as others'. Engels declared an enthusiasm for the novels of Eugène Sue, who wrote about Parisian life in the 1840s and 1850s, and often borrowed details of their artistic strengths (Messinger 1985, p. 59). Marc Blanchard's *In Search of the City: Engels, Baudelaire, Rimbaud* (1985) also raises the issue of Engels's embodied subjectivity in relation to his city walks and the definition of the urban experience in terms of a variety of aesthetically inflected discourses and perspectives.

In 1903 Georg Simmel suggested in his brilliant and brief essay "The Metropolis and Mental Life" that, for the sake of psychic survival, the metropolitan subject must relinquish the possibility of emotional life in city streets for what has become the very recognizable urban trait of a studied, nonchalant unconcern. His ideas regarding the cultivation of a blasé attitude as a defensive metropolitan mentality *contra* urban modernity marked his reception for most of the twentieth century (Simmel 1997). Simmel's full intellectual recuperation by the 1990s, begun by David Frisby the previous decade with *Sociological Impressionism: A Reassessment of Georg Simmel's Social Theory* (1981), reveals a thinker much more immersed than detached, his essays characterized by Frisby as "intellectual poems" (p. 76). Just as Baudelaire's, Simmel's work speaks to the socio-aesthetic approach to everyday life that has appeared in the last decade across the disciplines (Wolff 2000; Keohane 2002; Gerlach and Hamilton 2004; Highmore 2004; Cockburn 2005; Watts 2005; Jensen 2006). In "Assembling money and the senses: Revisiting George Simmel and the city," Michael Schillmeier links Simmel with actor–network theory (ANT) in an overdue account of the aesthetics of this "sociological Impressionist." Blanchard declares that he crosses Engels with Baudelaire and Rimbaud to produce what he has called a "third discourse, both creative and critical, spontaneous and reflective" (1993, p. 74).

Flânerie and globalizing cities

Susan Stanford Friedman (2005) writes of the shock of the futuristic area of Pudong across the Huangpu River from Shanghai in what had been paddy fields only ten years before and particularly of the jarring juxtapositions of the old and the new in this global city in the making. She declared that a new geography of modernity, a "polycentric" as opposed to "Eurocentric" modernity was emerging there, the key components of which include "the velocity, acceleration, and dynamism of shattering change across a wide spectrum of . . . intercultural contact zones." And while she did not mention emotional *flânerie* per se, she did say that what was bringing this new modernity into being was "a phenomenology of the new and the now: a self-reflexive, experiential dimension that includes a gamut of sensations from displacement, despair, and nostalgia to exhilaration, hope, and embrace of the new" (2005, p. 8). In *Walking Between Slums and Skyscrapers,* Tsung-Yi Michelle Huang does indeed designate *flânerie* as more vital than ever in the new global cities such as Shanghai, which she nominates as the ultimate habitat of walkers. In a chapter entitled "Walking in Shanghai," she superimposes social accounts of the reshaping of Shanghai's urban space, where one-fifth of the world's construction cranes "are in intensive service," with personal accounts of the city's walkers as they attempt to come to emotional terms with their new, everyday lived spaces.

The two authors speak to an increasing recognition among scholars of the need to chart emotional responses to the complex, provisional, transgressive, permeable, self-generating dynamics of today's globalizing cities. A Baudelariain inspired *flânerie* is regularly invoked to grapple with the urban kaleidoscope. There is now a more Baudelairean cast to contemporary *flânerie* as a practice of subjective mediation that establishes an ever-expanding, sensory connectivity among individuals in the streets, producing in the process vibrant documents of cities in transformation.

As the past gives way or mingles strangely with an ever-shifting present, *modern* surges back into scholarly discourse to describe the new shock effects of globalizing cities. *Interurban* replaces *international* as these new cities arrange themselves along networks that defy national boundaries, so that *cosmopolitanism* describes more of a reality than it ever has before. The fast-paced novelty of the urban experience once again demands direct, pedestrian immersion in the life of the streets for optimum interpretation (Figure 5.1). Thus, at the turn of the twenty-first century, *flânerie* was re-kindled as a subjective process involving walking around a city with all five of the senses recording at full bore, comprising aesthetic and social-scientific sensibilities alike, often intertwined. Artists began responding to today's globalizing cities by means of a *flânerie* reminiscent of that practiced by their nineteenth-century counterparts, responding to conditions and producing art that scholars have identified as belonging to a "Second Modernity."

Forms of embodied, affective Baudelairean *flânerie* also now inform much of the terminological currency associated with negotiating today's globalizing cities, such as:

- actor–network theory
- aesthetic journalism
- corporeal aesthetics
- embodied experience
- ethnographic mobility
- lived figuration
- reflexivity
- relational aesthetics
- rhythm analysis
- urban imaginary.

The terms are used to describe the effects of globalizing cities' rapid, radical transformation (Frisby 1988; Bauman 1990; Giddens 1991; Berman 1992; Jacobs 1993; Latour 1993; Featherstone et al. 1995; Pile 1996; King 1997; Nylund 2001). A perceived need for the reinvigoration of the modern, the embodied, the subjective, the reflexive, and the everyday shape these debates.

A significant outcome of the scholarly focus on the complex, provisional, transgressive, intersubjective, and encounter-driven dynamics of globalizing cities is the dawning realization that these dynamics require both ethnographic acuity *and* aesthetic sensitivity for their full effect. Urban theory's long-standing, primary focus on cities' economic, social, and political structures rather than on their sensual, spatial negotiation required overhaul: slow passage (preferably foot-based) through mundane minutiae, which produced a certain aesthetic rush, was the order of the day. The notion of "aesthetic reflexivity," as posited by Scott Lash, enabled aesthetic inroads into sociological inquiry (Beck et al. 1994; Lash 1999). Lash suggests that reflexivity, which charac-terizes the experiential passage through the messy everyday, becomes aesthetic as individuals' immediate, subjective responses to these passages "ground"

Figure 5.1 Walking in Baltimore.
Photo: John Rennie Short.

the disorder they encounter through the order of creative expression, which variously combines "community, history, tradition, the symbolic, place, the material, language, life-world, the gift, Sittlichkeit, the political, the religious, forms of life, memory, nature, the monument, the path, fecundity, the tale, habitus, the body" (1999, pp. 5–6). The Baudelairean kaleidoscope comes to mind. Lash speaks of the necessity for the "retrieval" of this multifarious ground as foundational of a "second modernity" that avoids the first modernity's universalizing and abstracting tendencies (1999, p. 6). Aesthetic reflexivity returns modernity to its phenomenological origins and in the process proves to be a valuable device for scholars to represent the diverse, sensual bombardment of globalization's urban cultures (Canclini and Chiappari 1995; Nowicka 2006).

The aesthetic urbanist meets the reflexive *flâneur*

The experience of the Late Modern Wave of Globalization revived *flânerie*. Mike Featherstone's (1998) important "The *Flâneur*, the City and Virtual Public Life," raised questions about how traditional ambulatory *flânerie* retains significance for the complexities of a globalizing city as well as how "electronic *flânerie*" might or should maintain any connection to the requisite physical embodiment. Anthropologist Joel Kahn (2003) noted that as the contemporary *flâneur* adheres to a "reflexive imperative" when confronting a global unity of "multiple modernities," s/he serves to conjoin the ethnographic and the modern.

Baudelairean *flânerie* used perambulation to amass knowledge of the Western metropolis as material for the production of journalistic, literary, or critical texts. Similarly, present-day *flânerie*'s translation of urban life from the direct, experiential contact of globalizing cities under development contributes to a socio-aesthetic production. Yet the twenty-first-century global *flâneur* is a new kind of aesthetic being with a decisive ethnographic bent, whose artistic practice has in turn been adapted by urban studies scholars. Today's *flânerie* is at once a research practice as art and art as research practice as it intertwines the approaches of social scientists and artists alike in a new form of interdisciplinarity (Pink et al. 2010). Examples of aesthetic urbanists include Bassett (2004), Butler (2006) Meagher (2007), Rao (2007) and Wunderlich (2008). A range of artist-*flâneurs* is shown in Table 5.1. The separate categories of aesthetic urbanists and artist-*flâneurs* are fluid. This interdisciplinarity is also a driving force in current exhibiting art collectives (see Table 5.2). Boundaries between fields that adapt *flânerie* do not just blur but disappear. *Flânerie* itself is a new discipline without borders—a tool of critical thinking *and* an art-making practice.

Today, the city motivates social imaginaries through urban poesis. This is the city as "a scene for multiple self projections, a space teeming with endless signification, a poly-palimpsestuous site inviting endless discoveries, mythopoeic activities in every direction" (Resina and Ingenschay 2003, p. 55).

Table 5.1 Artist-*flâneur*s

Name/nationality/medium	Project/date/exhibition or website	Place
Kinga Araya, Polish-Canadian, performance art	*Walking the Wall*, 2008: www.kingaaraya.com	Berlin
Claude Chuzel, French, video	*Work,* 2007: *Reconfiguring Cities* (Detroit, 2009)	Dubai
Hans Eijkelbloom, Dutch, photography/book art	*Paris–New York–Shanghai: A book about the past, present and (possibly) future capital of the world,* 2007	Global
Susan Hefuna, German-Egyptian, multimedia	*Cairo Crossroads,* 2008: www.susanhefuna.com	Cairo
Lin Ylin, Chinese, video	*Safely maneuvering across Lin He Road,* 1995 (Documenta XII, 2007)	Guangzhou
Elisabeth Neudörfl, German, photography	*Super Pussy Bangkok, 2006*: Von der Strasse (Galerie Barbara Wein, Berlin, 2009)	Bangkok
Song Dong, Chinese, performance, experience, video,installation	*Writing Time with Water,* 1995–2007: Tokyo, Beijing, Hong King, Lhasa, Venice, Chengdu, among other cities (Shanghai Zendai Museum of Modern Art, 2008)	Global
Shu Haolun, Chinese, film	*Nostalgia,* 2006 (documentary)	Shanghai
Stephen Schulz, German, video	*Equally Distant from Both Sides,*2008, www.maybevideo does.de	Montreal
Beat Streuli, Swiss, photography, video, installation	Individuals and crowds in city streets: New York, Osaka, Athens, Sharjah, Barcelona, Yokohama, among many others www.beatstreuli.com	Global

Flânerie's revival with a feminine focus

Evidence of the female urban stroller during *flânerie*'s first round is scant since a woman could not maintain the necessary incognito or enjoy the required free passage throughout the typical nineteenth-century city. There are exceptions: Karin Baumgartner, for example, presents a very convincing case for Helmina von Chézy, a correspondent for the journal *Französische Miscellen*, providing guidelines as early as 1803 on how to experience urban space as a woman (Baumgartner 2008). An important body of scholarly work

Table 5.2 Collectives of aesthetic urbanists and artist-*flâneurs*

Name	Purposes	Place
Creative Tools 4 Critical Times	Repository of urban/art projects; project funding; links to other collectives	World Wide Web
Museum as Hub	Project of the New Museum in New York to establish collaborative urban connections with hub partners throughout the world; exhibitions; exchange regarding national and urban issues	Cairo, Egypt; Eindhoven, the Netherlands, Mexico City, Mexico; Seoul, South Korea
Platform for Urban Investigation	Nomadic research laboratory for urban investigation, promoting cross-disciplinary, collaborative workshops in cities all over the world	Mexico, Eindhoven, Shanghai
Storefront for Art and Architecture	International exhibitions, events, archives, curated books	New York with transnational reach
Visible City: Project + Archive	Interventions in the social sphere, archives, publications and journals	Toronto, Helsinki, Havana

dating from the feminist priorities of the 1980s revises notions of the *flâneuse's* relative visibility in the nineteenth century. This scholarship asserts the *flâneuse* as a distinct reality in the nineteenth-century metropolis of the West, making a convincing case that public (so-called masculine) and domestic (so-called feminine) spheres were not so mutually exclusive as to completely preclude a proactive feminine presence in the streets. Just as her male counterpart, the *flâneuse* could, within certain limitations, achieve some anonymity on the street, be a detached observer, and produce social criticism and art from her experience. In *Streetwalking the Metropolis: Women, the City, and Modernity*, Deborah Parsons offers another, tantalizing scenario regarding the *flâneuse's* "absence" on nineteenth-century streets: she is actually hiding in plain sight as the feminine side of the typically dandified and androgynous *flâneur*; her incorporation informs his love of masquerade and fine-tuned, emotional receptivity to the city's sensations as well as reflects a growing male anxiety about the liberalization of public life (2000, p. 26).

Urban public life in the West became increasingly available to women by 1900. Yet just as a feminine filtration of the twentieth-century urban would seem to be imminent, the significance of *flânerie's* embodied process declined in the face of the perception that the managed city had lost its experiential immediacy and with it the chance of any gendered diversity. It has only been

Figure 5.2 Walking in Tunis.
Photo: John Rennie Short.

within the last two decades that the rapid and radical transformation in global-izing cities from South America to Africa through the Indian subcontinent and Asia has renewed *flânerie*'s credibility as an effective way through the unpredictable and transformative conditions inherent in today's urban spaces, rife with the creative chaos of self-generating networks of technological, economic, cultural, migrational, and even microbial flows (Figure 5.2). At the turn of the twenty-first century, *flânerie* regenerates as a sensuous, subjective process, and the examination of this reinvigoration in terms of the *flâneuse*'s urban experience is overdue.

Documentations of the dynamics of today's globalizing cities from the *flâneuse*'s perspective abound, twining the aesthetic and sociological threads of her experience into visual and/or verbal renditions, providing vivid docu-ments of cities in transformation. These accounts are complex, experiential, and emotive renditions of the urban experience, but many reflect the exurban condition created by twenty-first-century globalization. This may indicate women cannot yet confidently or consistently perform *flânerie* on some world stages and must remain at the peripheries, marginalized. However, this assumption may also be flawed: today's global *flâneuses* are not necessarily on the margins of *flânerie*, but rather at its frontiers as urban expansionism reinscribes the sites that newly accommodate walking practices. And it is just those walking practices that link *flâneuses* to traditional *flânerie* more than their venues.

The nomad *flâneur* and the globalizing city

So far, we have discussed a practice that draws upon visual culture and urban studies with the goal of positing *flânerie* as a cultural metric for globalizing cities, however porous their boundaries may be. We would now like to suggest that globalization's inherent interurban quality facilitates a new kind of *flânerie*, that of the global nomad, as another process to serve the experiencing, charting, and the conferring of "globalizing" to a city. We are not contending that nomad *flâneurs* are treading beyond even the exurban and walking across the globe, but they are moving along a global network, the burgeoning interurban circuit of contemporary art, comprising transnational biennial exhibitions and art fairs. A city's location along this network gives it a globalizing dimension. Cities barely recognized as major urban, much less art, centers a decade ago—Luanda, Sharjah, Guangzhou, for example—ring increasingly familiar as sites of contemporary art display and debate (*and* art making, more about which below), gaining degrees of cosmopolitanism in the process. And of course this is nothing new: mega-events such as World's Fairs and Olympics have longed served as vehicles to further global aspirations. Take the recently ended Expo 2010 in Shanghai—after it was over, Shanghai was even more firmly inserted into global flows of people and ideas. Now major exhibitions of contemporary art (biennials and triennials geometrically increase every year) similarly serve as dazzling vehicles for global aspirations.

What is clear in terms of the *flâneur* nomads is that they are born by a rising tide of cosmopolitanism, and this is also relevant to the more conventional form of *flânerie*. Long saddled with the negative connotation of a naïve, utopian Eurocentricism, cosmopolitanism has received renewed and positive attention in the light of globalization as something much more real than ever before. Baudelaire stood at the gates of Paris' 1855 Exposition Universelle and called for the "divine grace of cosmopolitanism" to descend upon a soul willing to be transported to China so as to "participate in the surroundings which have given birth to this singular flowering," in this case a "product of China," probably porcelain or perhaps silk (Baudelaire 1965, pp. 121–122). Today, this is realized not so mystically but phenomenologically in such traveling exhibition extravaganzas as *The Street Belongs to All of Us*, a project of the City on the Move Institute (Institut pour la Ville en Mouvement, based in Paris), which is a collective of artists, social scientists, architects, urban planners, and everyday people (www.ville-en-mouvement.com/uk/index.html). This collective has traveled to Buenos Aires, Santiago, Toronto, Beijing, and many other cities, in order to *flan* and create exhibitions unique to each place. Recently, the exhibition was in Shanghai in conjunction with the 2010 Shanghai Biennale, whose theme was "Better Mobility, Better Life" (mirroring the concurrent Expo Shanghai's theme of "Better City, Better Life"). Here is an example of transnational mobility resulting in urban *flânerie*. It is also an example of what Beck and Sznaider (2006) call the "Cosmopolitan Condition," which they describe as the result of the dissolution of the dualities of the global and the local, the national and the international, us and them.

The nomad *flâneur* practices a broader, more cosmopolitan form of *flânerie* than the strictly urban variety and blends the fractures of global society and urban polarization while still keeping their narrative balance. The charge to the global nomad *flâneur* is to provide meaning for the space between the global and the urban/rural, to make cosmopolitan sense of the global metropolitan. In the twenty-first century, cities are new and strange again as reterritorialized modernity in globalizing cities, particularly beyond the West, combine with global and local characteristics to provide cityscapes with quotidian yet phantasmagoric experience for the *flâneur* to shape.

Endings and beginnings

Flânerie is alive and well. In response to the rapidly growing globalizing cities, especially in the developing world, both its aesthetic and social strands are being renewed. Complex, experiential, and emotive documentations of the dynamics of today's world cities from the *flâneur*'s perspective provide not only vivid documents of cities in transformation but also represent the new urban imaginaries at their megalopolitan peripheries and beyond.

Flânerie did not die out with the commodification and commercialization of modern culture, as Benjamin said it did, but neither did it remain untouched. The contemporary *flâneur* is recast in a new urban order. With the globalizing city as a theatrical setting for the project of globalization, the stage for globalizing, the modern *flâneur* is witness and conspirator in a new global urban narrative. From nineteenth century to twenty-first century the *flâneur* moves from dispassionate distance to co-conspirator in new urban meta-narratives. In the nineteenth century it was the *flâneur* that donned the carnival mask, in the twenty-first century it was the city. The old *flâneur* was always engaged in a battle against the phantasmagoria of commodities that he saw as threatening to the practice of strolling throughout the city: he saw it as a distraction and saw himself as an endangered species because of it. He battled it by trying to maintain a certain psychic distance from it, by maintaining a certain superiority to it, by resisting the lure of commodification. The new *flâneur* submits to the commodification. The original *flâneur* confronted and rearranged urban disorder in a narrative of free floating signifiers; he was a producer, author, and detective. The new *flâneur* searches for personal disorder in the new urban order in a narrative of marked signifiers; he/she is a consumer, a visitor, a tourist, a fan (Figure 5.3).

A number of questions remain: First, can women confidently or consistently perform *flânerie* anywhere? Second, do other forms of mobility and communication undermine the traditional notion of *flânerie*? Third, there is the

Figure 5.3 Celebratory street life in Warsaw.
Photo: John Rennie Short.

problematic nature of the knowing city. One of *flânerie*'s original roles was to produce a creative reading of the city through a rather random traversal. This has been made more problematic because the city has become more self-conscious and explicitly legible. This takes a variety of forms from the identity of heritage routes, protected neighborhoods, tours of the downtown, and the identity of "ethnic" neighborhoods. So, for example, Chinatown has become less a foreign enclave than a commodified experience. This is certainly true of cities in the West where *flânerie* originated. The original *flâneur*'s aimless yet creative wandering produces meaning, largely through individual autonomy: Can this still be possible today? We can make the claim that the Baudelairean experience of the early *flâneur* is still possible, especially in the emerging cities of the developing world. Engels described the cities of nineteenth-century England as "shock cities": they are like a foreign body emerging from society. The shock cities of the twenty-first century are in Latin America, Asia, and Africa, growing quickly, being remade and restructured at dizzying rates. The *flâneurs* of today (whatever they will be called) will emerge from this context.

There are also continuances. Modern *flânerie* is immersed in labyrinths of the urban, extra-urban, interurban, nomadic, and diasporic. The pedestrian experiences of the contemporary *flâneur/flâneuse* would shatter Baudelaire's kaleidoscope, yet they maintain strong ties to classic *flânerie* in that they translate their wanderings into a combination of both art and social science. As a mode of analysis combining aesthetic, geographic, and ethnographic attention to the urban experience, the complexity of globalizing cities requires nothing less.

CASE STUDY 5.1 **The comfort of the grid**

In May–June in 2009 I was staying in Milwaukee, Wisconsin. I was working at the American Geographic Society Map Library at the University. My daily journey to work was a fifteen-minute walk from an apartment complex to the university library. It was an easy path to chart as I followed the rigid grid system of the street layout. Now the concept of the grid has come in for a lot of criticism over the years, particularly regarding the notion of an "unfortunate" triumph of geometry over geography, of an artificial human-made network with unforeseen conse-quences imposed on the natural landscape. And yet the grid system also has its advantages: it is a great comfort to visitors in a new city, easy to understand and simple to navigate. I contrast such ease with my experience in Tunis in August 2008, when I wandered through the tortuous street layout of the old Medina, hopelessly lost and pulling a suitcase through the crowded, narrow walkways. Arabic curses filled the air behind me as my suitcase bumped against unsuspecting legs. Disorientated, sweating profusely, and on the verge of panic, I reached my hotel more by sheer luck than guided skill. In contrast to the unknowable layout of this centuries-old organic street pattern, the grid is a friend to the visiting stranger.

My commute in Milwaukee was along North Prospect Avenue through a neighborhood of dwellings built as the nineteenth century turned into the twentieth. What I like about this area is the quality and quantity of usable public space. From the edge of most houses, there is a porch that extends about 10 feet, an unfenced front garden of about 30 feet and then a sidewalk, of around 20 feet. The early summery weather coaxed people onto their porches and into their gardens. The porches and gardens are semi-public spaces, as people using them become part of the total street scene. The street reminded me of a remark by Jane Jacobs, the grande dame of Urban Studies, who noted that the more people we have on the street, the safer we feel. This neighborhood contrasts with many of the new upmarket areas, where despite the sinuosity of the street pattern (a contemporary reaction to the grid), the amount of public space is minimal; many affluent areas have no sidewalks, and few have the generous public and semi-public space along North Prospect.

We need more usable public space. It is not police that make us feel safe in public but the presence of lots of people going about their everyday business. Other people make us feel safe. And to use public space we have to be generous in our allocation of usable public space. We need to avoid design determinism; forms of design facilitate rather than cause human behavior. But the quality and quantity of public space does have an effect on the livability and conviviality of cities.

CASE STUDY 5.2 **Walking the High Line**

In 1934 an elevated rail line opened in Manhattan. It was a freight line designed to deliver goods and material to factories and warehouses along its route. It snaked roughly 13 miles north–south through Chelsea and the Meatpacking district in the southwest corner of the island. As freight shifted to truck rather than rail and as the warehouses and factories closed, the rail line was no longer viable. The last delivery was made in 1980. The narrow railway lay empty, used by the occasional daring urban explorer. There were plans for demolition. In 1999 the Regional Plan Association, commissioned by the railway line owners CSX Railroad, suggested turning the line into a pedestrian promenade. A non-profit organization, the Friends of the High Line, was soon established, raised money, and held a design competition. In 2009 the first section of the linear pedestrian park was opened. Native plant species were planted along its route, growing up beside the old rail tracks.

Figure 5.4 Walking the High Line in New York City.
Photo: John Rennie Short.

The High Line was the second elevated railroad turned pedestrian walkway. The first was the Promenade plantée in Paris. The High Line quickly became popular, attracting both locals and visitors. It gives a wonderful walking experience looking down at the hip neighborhoods. The park stimulated other projects. The Whitney Museum is planning to open a new museum at one of the entrances. The Standard Hotel, which straddles the High Line, has floor to ceiling windows that give lots of opportunities for exhibitionism and voyeurism that were long one of the charms of flânerie. The High Line hosts dance performances and musical events. Innovative artwork is exhibited along the route. It is now an immensely popular and well-loved public space. It also appears in television adverts.

The conversion of an abandoned railroad track is just one example of the revival of a managed *flânerie*. The postindustrial late modern city has its opportunities for the transformation of industrial space into leisure space as the dominance of the machine is replaced by the lure of human locomotion. The creation of an urban public space for walking is one of the more successful transformations (Figure 5.4).

The official website for the High Line is www.thehighline.org/.

CASE STUDY 5.3 **The phoneur**

You see them everywhere, walking or driving, a phone is clutched to their ear or a small screen is in front of their eyes (Figure 5.5). Often their conversations are louder than ordinary speech. We quickly assess the relationship and its trajectory from the sometimes all-too-loud conversation. Or else they are busy texting or surfing the web, their fingers moving quickly, the eyes scanning the screen as they walk and sometimes talk.

The attention of the original *flâneur* was on the here and now. The late modern *flâneur*, now technologically connected, is what Robert Luke (2005) refers to as the "phoneur," someone whose attention is only partially in the here and now. The hereness and the nowness of the immediate embodied experience now mediated through a connection with the web or a conversation with someone else somewhere else.

Is the phoneur just a *flâneur* with a mobile phone? A source of liberation, the phone, especially one with internet connections, allows the strolling of the cityscape with more information quite literally at hand more than the immediate sensory experiences: hungry, look up a list of good restaurants; wanting something to buy, look up a consumer site. The phoneur walks the streets and surfs the web, walks though the city but stalks cyberspace. Free and connected, the phoneur has the technologies of empowerment as events and actions by authorities are witnessed, recorded, and transmitted. The students protesting the repression in Iran come to mind as they documented police brutality.

Figure 5.5 A phoneur in Barcelona.
Photo: John Rennie Short.

There are some who question how the phoneur is incorporated into a commodity culture and surveillance society while others point to the liberation aspects (Ibrahim 2007). Whatever the mix of commodification and liberation, the phoneur signifies a new relationship between people and place. The very sense of place is now mediated through a widening and deepening technological membrane that connects and disconnects as it liberates and constrains.

---

6 Traffic, accidents and the modern city

At the heart of modernity is the connection between speed, the machine, and the city. In his 1909 *Manifesto of Futurism*, Filippo Tommaso Marinetti (1876–1944) glorified the man at the wheel and the beauty of speed. He associated speed with courage in action and slowness with stagnant prudence. His glorification of the speeding automobile in the city prefigures its subsequent centrality to urban development. Cities were reimagined and then re-engineered to promote high speed. Virilio (1986, 1995, and 2005) tells a story of the destruction of the sociability of the city by the logic of acceleration. In a more nuanced interpretation, Latham and McCormack (2008) explore the ways in which speed and the "countervailing eddies of slowness" define the experience of the city. However, in terms of urban traffic and the marginalization and displacement of the self-propelled human body, Virilio's image of the city as a site of acceleration seems closer to the mark.

The modern city is designed largely around the use of motor vehicles despite the inherent risks in machine–body coexistence, part of the brisk pace of city life. Indeed, in the immediate prehistory of automotive ascendancy, the Futurists downplayed any risks and celebrated machine–body meshing as a means to a high-speed, dynamic superhumanity. Such eventual transformation was made possible, according to Futurism's founding myth, when poet Filippo Marinetti emerged from a ditch after crashing his beloved Fiat, spewing mud and apparently, to quote the *Manifesto of Futurism*, which begins "We want to sing the love of danger, the habit of energy and rashness," famously continues:

> We declare that the splendor of the world has been enriched by a new beauty: the beauty of speed. A racing automobile with its bonnet adorned with great tubes like serpents with explosive breath . . . a roaring motor car which seems to run on machine-gun fire, is more beautiful than the Victory of Samothrace.
>
> (Flint 1972, p. 41)

By the early 1930s, Le Corbusier further enshrined swift automobility, proposing his "autostrades" as concrete ribbons threading skyscrapers'

Traffic, accidents and the modern city   143

canopies, elevated above mere city streets and devoid of pedestrians for optimum vehicular velocity. Speed and the fantasy of infinite mobility were at the heart of modernism and especially as its realization in urban modernity.

Peter Norton (2008) tells the story of US cities: at the beginning of the twentieth century, city streets had a variety of users—people as well as cars— and multiple uses, a place for children playing, people walking, neighbors chatting. From the 1900s to the early 1930s, a battle was fought between, on the one hand, the multiple users and interests decrying the onslaught of cars on streets and the growing dominance of auto traffic—the term "death cars" was frequently used—and on the other hand, automotive interests that continually invoked freedom as a rallying cry. The automotive interests won. From our perspective today, it seems a foregone conclusion, but the twenty-five year period in the early twentieth century reminds us of the battle for alternative conceptions about the primary purpose of city streets.

This urban restructuring of city streets as pathways for automobiles has a high human cost. High-speed machines mangle bodies and kill pedestrians, and yet these costs are largely absent from urban studies and discussions of urban safety. *The Safe City* (Berg et al. 2006) makes for interesting reading, as much for what it does not, as for what it does, include. While issues of terrorism and crime abound in the book, the risk of bodily injury and fatality from traffic accidents receives scant attention despite the book's title. Such neglect is not uncommon in the recent urban literature where terrorist campaigns and crime waves dominate safety issues while there is a huge silence about traffic safety and especially about pedestrian injuries and fatalities. Numerous commentators have drawn attention to traffic and pedestrian safety (Hass-Klau 1990; Hillman 1993) as have numerous empirical studies. However, the issue has failed to gain traction in the wider and broader urban studies literature, despite its obvious importance. Across the world nearly ten million people are crippled or injured each year, and approximately 1.2 million people are killed every year due to road accidents, approximately 3,250 people every day (Peden et al. 2004). That is the equivalent of a World Trade Center disaster on the world's roads each and every day. One writer even refers to a global epidemic of road deaths (Dahl 2004). The total economic cost is estimated at 1 percent of the GNP of low-income countries and 2 percent in high-income countries (Jacobs et al. 2000).

No accident

Why the silence? Two reasons stand out. First, there is the notion of the accidental, of random events that are to be endured rather than explained. The term "accident" is revealing: it initially meant an event, anything that happens. Now the word commonly refers to an *unforeseen* event, often a mishap. But the predictable covariance of road deaths with social phenomena belies this connotation of "accident." The young and the vulnerable, the poor and the marginal, are more likely to be hit by a car than the rich and the powerful.

While road deaths have declined in rich countries, they have increased in poor countries. These statistics suggest an ordered causation rather than random occurrences. Road traffic accidents are not mishaps but events that are both explainable and preventable. The relative silence shrouding this serious issue is itself a political act since the incidence of pedestrian fatalities and injuries is skewed toward the marginal and vulnerable members of society. Yet this consistent finding has failed to capture the attention of most critical theorists. Most pedestrian fatalities occur in poor countries and have greater impact on the poor, the young and the old. Pedestrian injuries are just one strand in webs of multiple deprivations that bind disadvantaged people. To recognize traffic fatalities and injuries through marginalization theory is to offer a very different perspective on them as "accidents."

Second, there is the silence assigned to all events that have a constant background quality. The fact that road traffic accidents happen all the time blends them into the fabric of the "taken for granted." Too regular to generate much notice, they become mere white noise rather than events to be analyzed. Their very consistency and regularity allows them to fade into an unexamined, rarely discussed, space. The urban spectaculars such as the collapse of the World Trade Center, overwhelm and ultimately silence the murderous regularity of road traffic fatalities and injuries. Road traffic "accidents" are a taken-for-granted seemingly inevitable cost of contemporary urban living: too mundane, too anonymous, too ordinary to generate much interest. And yet the rising concern with the everyday ordinary nature of the lived urban experience should prompt a reconsideration of road safety and risk.

The "ordinary" quality assigned to road traffic accidents is, if you excuse the unintended pun, no accident: there is also a vast array of interests concerned with making them ordinary. These interests include the motor vehicle manufacturers, road builders and construction companies as well as a society that does not want to question the fatal costs of a culture and of cities dominated by fast-moving vehicles (Figure 6.1). The current usage of the term "accidents" to refer to fatal and damaging machine–body interactions (especially in the city) is incorrect, misleading, and deceptive. The usage shifts the phenomenon from social outcome to random event. Charles Perrow (1999) introduced the notion of "normal accidents" that may be unforeseeable yet are inevitable. Perrow based his initial work on the near meltdown of Three Mile Island. Drawing a wider perspective on high-risk technologies, he argues that accidents occur when systems become more complex and interconnected. These conditions are increasingly met in cities where traffic patterns are more complex and a variety of users share busier roads. The ubiquitous nature of motorized transport in many cities, in comparison to the obvious singularity of the nuclear power station renders it, in the minds of many, as safe or relatively innocuous. Yet a heavy piece of metal—the average weight of a US car in 2006 was 4,142 lbs. (1,878 kg)—traveling at even 30 mph, operated by someone, perhaps listening to the radio, drinking coffee, or using their

Figure 6.1 Car traffic dominates streets in Naples.
Photo: John Rennie Short.

Figure 6.2 Pedestrian navigates traffic in Tunis.
Photo: John Rennie Short.

cellphone, times the thousands of other drivers on a road in the average city, constitutes a complex technological system with lots of room for human error. Our need for speed and greater reliance on urban vehicular traffic has in part blinded us to its inherent and inevitable risks to health and safety. Popular perceptions of risks, in general and in relation to traffic accident risk in particular, are based more on intuition and judgment, rather than objective assessments. Traffic fatalities and injuries are rarely discussed as part of the human costs of space–time convergence and space–time compression.

As the more affluent move around by car and limousine, the pedestrian becomes the minor player in transport discussions, and as cities are restructured to fit the needs of the car, then the needs of the pedestrian are even less considered (Figure 6.2). But even the car driver has to step out of the car at times; while not everyone drives a car, most of us are pedestrians at some time. However, as soon you step out of your car in many if not most cities, you become a second-class citizen: it is as if your full rights only exist when you drive a car. Walk in the city, and you are pushed to the sides of the road, a telling phrase and metaphor. When a car hits a child it is often treated as an accident, when a child hits a car it is considered vandalism. When I wrote in *The Humane City* of cities designed as if only some people matter, I referenced specifically car drivers in contrast with pedestrians (Short 1989). Cities are biased toward the needs of the drivers rather than the rights of the pedestrians, despite the obvious inequality. A heavy metal object traveling at even relatively low speeds does more damage to a human body than the body does to the car. People get killed and maimed; cars merely get scratched and dented.

A wider theoretical context

There is an emerging literature on mobility. A key paper by Sheller and Urry (2006) identifies a new mobility paradigm that is undermining the sedentary nature of traditional social science theories. They draw attention to the importance of multiple mobilities and new dematerialized mobilities. There is now a growing emphasis on movement rather than stasis and on fluidity rather than structure; a shift marked by relational notions of the "space of flows" (Castells 1996) and embodied in the title of recent works such as "liquid life" (Bauman 2005) and the "liquid city" (Short 2007).

The notion of automobility is part of this "mobility turn," recognition that the automobile plays a central role in social life, identity and mobility (Featherstone et al. 2005). Mobility scholarship draws attention to the pervasive culture of the car, but tends to downplay the competing and complementary role of walking. Walking is still very much associated in popular imagination with images of Romantic poets and transcendentalist philosophers —an important element of the contact with nature but little studied in the current sociological literature as an important form of mobility or identity in

Figure 6.3 Car dominated cities create sterile spaces reserved for machines: Laurel, Maryland.

Photo: John Rennie Short.

the contemporary city. The project of modernity has sidelined the premodern reliance on human locomotion as a mode of travel and as a valuable perspective. The new mobility and automobility literature is imbued still with the emancipatory, a modernist legacy that tends to celebrate the ease and speed of movement and attends less to the notion of competing mobilities in the juxtaposition of automobility and walkability. And, as a consequence, even less attention is paid to the discussion of automobility's negative affects for pedestrians. The rise of the car is at the expense of the pedestrian. The rise of automobility is at the rising costs and danger of walkability.

A number of more general themes inform our concerns. The most important is the general notion of space–time activity, as first outlined by Hägerstrand (1970) who formulated space–time paths shaped around pegs, which structure our day, and prisms, which define accessible spaces available in specific time periods. Hägerstrand's framework allows us to imagine more clearly the city as a place of people and machines sharing space–time paths. Normark (2007) echoes this theme in the concept of "enacting mobility." Following a space–time path through the city has a different drama whether you are a walker or driver. Engwicht (1999) characterizes city streets as dual spaces for movement and exchange. Engwicht's basic belief, and one that we endorse, is that there are too many movement spaces for cars and not enough exchange spaces for pedestrians (Figure 6.3).

Traffic "accidents," the city, and modernization

In cities in the developed world, with their longer history of modernization, accidents have been reduced by subtle changes made over the years. Long-term adjustments, such as better traffic management, more stringent legislation, such as bans on drinking and driving and the restructuring of the city to marginalize pedestrians, have produced a decline in fatalities and injuries. However, in cities across the developing world, there is rapid modernization associated with the more recent enthronement of traffic mobility without the long-term adjustments that reduce "accidents." The result is an increased rate of pedestrian fatalities and injury.

There are 1.195 million deaths caused each year by road traffic accidents, and they constitute 6.5 percent of preventable deaths. In 2002 road traffic accidents caused 1.5 percent of deaths of children under fifteen years old and 5.0 percent of deaths of adults aged fifteen to fifty-nine years. Road traffic accidents injure between twenty and fifty million each year, and the estimated costs are $518 billion in medical care and $326 billion in productivity losses (Jacobs et al. 2000). The estimated costs to low- and middle-income countries amount to US$65 billion, more than the amount of money received in all forms of aid.

While traffic deaths have increased across the globe, from 990,000 in 1990 to nearly 1.2 million in 2002, there has been a marked spatial shift. Fatalities have declined in the developed world due to better vehicle design, improvements in safety standards such as drunk driving enforcement, higher seatbelt usage, better traffic management schemes, and the implementation of road safety programs. The decline is also the result of changes in pedestrian behavior in the city. The unforeseen consequences are new health risks. While people and especially children may avoid traffic accidents, it is because they are not playing in the street, and their recreation is now more sedentary. Reduced traffic fatalities and accidents are often purchased at the cost of other negative health outcomes. Neighborhoods and cities built around the automobile promote a more sedentary lifestyle, which is associated with the development of various health conditions, including heart disease, obesity, diabetes, asthma, and mental illness (Kawachi and Berkman 2003). Berke et al. (2007) argue that overweight and obesity are associated with non-walkable neighborhoods. When roads become too dangerous for children to play and for pedestrians to use, physical activity is then limited to the inside of the homes and the associated negative health outcomes emerge (Farley et al. 2007). This limitation in physical activity and prolonged exposure to various indoor pollutants can have significant and powerful health conse-quences. There is now a considerable body of literature that suggests a clear link between urban form and obesity levels (Lopez 2004; Krisberg 2006). A car-orientated urban sprawl is part of the mix of factors creating the obesity epidemic, particularly in children, in part, because keeping children safe from traffic means taking them off the streets. Physical activity is thereby reduced. This is a complex situation and more detailed research is necessary to tease out the specific explanatory threads between urban forms that prioritize vehicular traffic, thus restricting everyday physical activity that in turn can lead to increased obesity, especially among children.

In the US traffic accident fatalities have been falling and are now at their lowest level in three decades. The pattern, replicated in other rich countries, is of a steady increase in the number of fatalities throughout the twentieth century as vehicular traffic increased, peaking at over 55,000 deaths per year by the later 1960s. By 2000 the rate of fatalities per vehicle miles of travel declined from approximately eleven per 100 million vehicle miles of travel to just less than two per 100 million. The absolute number of fatalities is still high in the US. In 2003 there was a total of 42,643 deaths and 2.89 million injuries, and the total estimated costs amounted to $230 billion or 2.3 percent of the national GNP. The death rate is the equivalent of two fully loaded 747 jets crashing into each other every week. Imagine the media coverage and the academic rush to print in the wake of this pattern of events. A total of over three million people were killed on US roads in the twentieth century. Road traffic crashes are still the leading cause of injury-related death and the leading cause of death for persons aged one to thirty-four. Traffic safety programs receive only one percent of the US Department of Transportation budget.

While road traffic deaths and injuries are declining in developed countries, they are on the rise in developing countries. From 1975 to 1998, Canada's fatality rate declined by 63 percent, while China's increased by 234 percent; fatality rates declined by 27 percent in the US but increased by 383 percent in Botswana (Peden et al. 2004). The death and injury rates in low- and middle-income countries are probably much higher due to endemic underreporting. Africa is a continent with only 4 percent of global vehicle registrations yet 10 percent of global road traffic deaths. The injuries and fatalities are concentrated in the cities with their rapidly expanding use of vehicles yet much slower adoption of traffic safety measures.

The escalation in traffic deaths and injuries in poorer countries is due to a rapid increase in motorized vehicle traffic without a corresponding rise in safety improvements. There is a growing disparity between increases in vehicular traffic and the ability to manage this traffic safely. Coping with increased vehicles in cities is a learned experience. It took almost eighty years for the developed countries to effectively reduce the rate of fatalities and injuries. New traffic layouts had to be tested, planned, and built, new behaviors had to be learned, and new norms of car driving and pedestrian behavior had to be established, maintained, and ingrained. All this takes time and money. Disparities in wealth feed directly into spending on road safety. While a high-income country such as the United Kingdom spends nearly $39.00 (in 1980 US dollars) per capita on road safety, Pakistan and Uganda spend $0.07 and $0.09 per capita, respectively (Bishai et al. 2003).

Most developing countries have seen a very sharp increase in road fatalities and injuries. The major reason is rapid motorization without a corresponding increase in road safety. In Vietnam, for example, the motorization is largely the result of a significant increase in two-wheeled motorcycles. There were four million of these vehicles registered in 1995 but fourteen million by 2004. In the same period road traffic fatalities increased five times. In Hanoi, for example, roads are very narrow; motorcycles and pedestrian often share the same public space; speed limits are often broken; stop signs and pedestrian crossings are rarely adhered to; taxis and cars do not have seat belts and child car seats are not available. Motorcycles are often overloaded with multiple family members, and child passengers rarely use crash helmets. Across the country approximately forty people die every day in traffic accidents and almost eighty a day suffer from debilitating head injuries. While the global mortality rate due to traffic accidents was nineteen per 100,000 population, this figure was twenty-seven per 100,000 for Vietnam (Peden et al. 2004). Over 4,100 children die every year in Vietnam from traffic accidents, almost eleven children a day.

Pedestrians are particularly vulnerable in cities of developing countries. While pedestrians account for approximately 12 percent of road traffic deaths in the US, they constitute 50 percent, 45 percent and 35 percent in Ethiopia, Kenya and Malawi respectively. In Delhi, India, 55 percent of road fatalities are pedestrians and bicyclists (Mohan 2002). While most fatalities in the developed world are passengers in vehicles, most fatalities in developing countries are pedestrians, bicyclists, and public transport passengers, especially those in multi-passenger vehicles such as buses, that, in turn, lead to fatalities. While 10,000 crashes in the US lead to sixty-six deaths, the same number in Kenya results in 1,786 and 3,181 in Vietnam. Because of poorer medical provision, crashes are more likely to lead to fatalities than in the more affluent countries. There is often limited and sometimes no effective emergency medical response. Again, there is an emphasis on the dramatic: while medical provision and aid from the rich world is often available for the spectacular and highly televised event—the earthquake, the hurricane, the tsunami—the steady slaughter on the roads fails to evoke such generous responses.

To summarize the global distribution of traffic accidents, traffic fatalities and injuries are increasing at a global aggregate level, although there has been a marked decline in the rich world. Since 1979, traffic fatalities have declined 50 percent in Canada, 46 percent in Great Britain, 48 percent in Australia, and 18 percent in the United States (Evans 2003). In contrast, fatalities and injuries are on the increase in low- and middle-income countries where a surge in vehicular traffic is not matched by corresponding improvements in safety or medical provision. Nine out of every ten traffic fatalities occur in low- and middle-income countries where the most vulnerable are pedestrians, cyclists, and public transport users. Among pedestrians, the poor, the young, and the old are at especially high risk.

Pedestrians in the modern city

Worldwide, more than one-third of all traffic-related deaths result from pedestrian–vehicle crashes. Child traffic injuries are a major health risk in urban areas around the world in rich as well as poor countries (Durkin et al. 1999). Hewson (2004) argues that the UK has one of the worst records among European nations in child pedestrian safety. He highlights the fact that neighborhood deprivation has a negative effect on safety. White et al. (2000) show with reference to Scotland that the risk of death for child pedestrians is related to socio-economic status. They show that children from socio-economically disadvantaged families have a higher risk of physical injury and their injuries are more severe, and that the risk of pedestrian injury is over 50 percent higher for the children of single mothers. This statistic reinforces the finding that accidents are not random across the population.

In the US, on average, 6,000 pedestrians are killed each year and 90,000 are injured. A pedestrian or bicyclist is killed every four minutes. About a quarter of all traffic-related pedestrian deaths occur among children up to age fourteen. Among children aged from five through fourteen, the motor-vehicle crash death rate is more than five times that of drowning, the second most common cause of unintentional injury. About half of all deaths among adolescents occur as a result of motor-vehicle crashes, with a death rate two times higher than the second most common cause of external injury (U.S. Department of Health and Human Services, Health Resources and

Services Administration, Maternal and Child Health Bureau 2005). Grossman (2000) identifies several risk factors for traffic-related deaths describing an important link between the urban environment and human vulnerability. Living in neighborhoods with high traffic volume, few playgrounds, and few safe crossings increases the risk of being hit by a moving vehicle.

In Mexico, for example, motor-vehicle collisions represent the majority of the cases for morbidity and mortality. The apparent inability of pedestrians to safely cross intersections is a main contributor to injury, especially where traffic signals are exclusively for drivers and in some cases no bridges or lights are available to pedestrians. Boys with siblings who play on the streets for long hours also have higher risks for traffic injury, particularly in urban settings (Celis et al. 2003). There is a decreasing tendency for pedestrian injury in Mexican adolescents; probably because of less street play than in previous generations and a decrease in general physical activity.

Urban transport debates, reflecting the modern legacy, are still dominated by performative criteria; the overriding concerns revolve around questions of how we can make traffic move more efficiently, more quickly, through the urban environment. When safety is an issue, the emphasis is toward improving the safety of drivers and passengers rather than pedestrians. The emphasis tends to be on changing pedestrian behavior with the underlying assumption that pedestrians *must* be trained to acclimate to a car-dominated culture. Such policies and practices encourage a vehicle-centered perspective, where vehicles are given the power, control, and authority, leaving pedestrians at the mercy of drivers, even when limits are placed on vehicular traffic. Little evidence exists on pedestrian education effectiveness whereas urban traffic modification has been shown to be more effective (Richter et al. 2005). Freeman et al. (2004) evaluated several community-based interventions on unintentional injury prevention in the US, focusing heavily on educational interventions targeting school-age children, and found little benefit in reducing the incidence of fatal pedestrian injury. A much more effective intervention is educating vehicle drivers on the needs and rights of pedestrians. Vehicle design considerations reduce pedestrian fatalities by 20 percent when they focus more on the effects on pedestrians rather than on vehicle occupants (Crandall et al. 2002; DiMaggio et al. 2006).

There was a tight nexus between urban design/planning and public health during much of the nineteenth century when the problems of the new industrial city created many of the public health practices and institutions that still exist today. The emphasis on clean water, better-designed houses and neighborhoods all grew out of public health and urban planning principles forged in the rapidly growing and industrializing cities. The contagion effect of disease that affected the rich and powerful as well as the poor and weak prompted citywide public policies. However, by the late nineteenth and through most of the twentieth centuries, traffic deaths and injuries were not seen as a public health issue in the same way and did not prompt the previous level of strong legislative advocacy. Traffic was seen primarily as an economic issue, not an urban health issue. Pedestrians were the cause of accidents and thus needed to be educated and disciplined.

New frameworks

A preventative framework can be built around three urban design/planning issues that have an important effect on reducing the risk of pedestrian injuries. The first is reducing the volume of traffic in urban neighborhoods. Traffic volume reduction has the potential to reduce incidence of injury. The second focuses on environmental modifications that separate vehicular traffic from pedestrians. The use of traffic signals, under/overpasses, wide sidewalks, refuge islands, and the installation of roundabouts and multi-way stop signs reduce pedestrian crashes by 75 percent. Retting et al. (2003), for example, demonstrate that modification of the built environment to enhance vehicle–pedestrian seperation and to increase the visibility of pedestrians reduces the risk of vehicle–pedestrian crashes. While there are many small-scale changes that can make cities more pedestrian friendly—increased pedestrian visibility and conspicuity, better lighting, parking restrictions, bus stop relocation—ultimately, of course, this requires a reorientation of our cities so that urban neighborhoods are seen more as places to live in rather than as places to drive through. The third is reducing vehicular speed. A major cause of injury is the high speed of vehicular traffic. The lack of speed restrictions causes child traffic deaths even more than increased exposure to traffic. Many pedestrians are killed and seriously injured because cars and trucks are traveling too fast. Havant Borough Council in the UK saw a casualty drop of 40 percent when a 20 mph speed limit was imposed on 20 miles of road (Pilkington 2000). In the state of Maryland, in the US, one-third of the 651 vehicle-related fatalities in 2006 were caused by speeding. When one of the counties, Montgomery County, adopted speed cameras, the number of drivers going 10 percent over posted limits dropped by 70 percent. Increased traffic volume and vehicle speeds of 40 mph and greater are correlated with an exponential increase in injury risk (Mohan et al. 2006). The simple message: speed kills and vehicular speed kills pedestrians.

A combination of speed limit reductions, traffic-calming measures, driver education, and road engineering would help achieve long-term reductions in pedestrian fatalities. In the past, emphasis was placed on effecting change in pedestrian behavior, making them conform to the needs of fast moving vehicles through urban streets. A more radical shift involves recreating cities that privilege pedestrians over drivers.

The city as urban walkshed

On September 13, 1899 Henry Bliss stepped off a streetcar in Manhattan and was hit by a taxicab. He died the following morning. He was the first recorded vehicle fatality in the Americas. The relationship between pedestrians and drivers of vehicles has always been uneasy. As vehicular traffic increases, so do the risks to pedestrians. Attempts to regulate drivers and pedestrians date back to the nineteenth century. Traffic signals were first installed in 1868 in the UK. In 1890 it was first suggested that under/overpasses should be built for pedestrians. All these safety measures took second place, however, to concerns with ensuring the free flow of traffic (Ishaque and Noland 2006). In the popular US magazine, *The Saturday Evening Post*, as long ago as 1949, a journalist noted:

> The pedestrian is much like a hunted animal always in season. Armored only by thin layers of clothing and his own tender hide, he is being slaughtered at the rate of about thirty a day and mangled at the rate of nearly 700 a day.
>
> (Wittels 1949)

In the twentieth century the city was restructured to prioritize vehicular traffic. Streets lost their conviviality for pedestrians as they were engineered primarily as routes for speeding traffic. In some small places the tide has begun to turn, especially in the developed world. Pedestrianization schemes are proving popular with city residents. Some streets are being remade so that traffic speed and movement is minimized. Traffic-calming techniques are being introduced and the design principles of the new urbanism call for more walkable streets. In numerous suburban subdivisions, traffic-calming measures are built into street layouts of curvilinear lines and cul-de-sacs with an emphasis on safety rather providing a grid of perpetual vehicular movement. The notion of a slow city is taking root. Mara Miele (2008) describes the international network of CittaSlow, small towns devoted to producing slowness. It originated in Italy. There are around 100 slow cities worldwide with more than half in Italy. While the emphasis is on slow food, I find the overall concept intriguing, purposefully slowing down the pace of the city.

We can also reimagine the city as a series of urban walksheds. A walkshed is the space–time cone of walkability centered at the primary residence or place of work or play. Picturing the city as a series of overlapping walksheds is an important start to understanding, conceptualizing, and mapping the city as a convivial place for pedestrians. Measuring the safety and quality of life within a walkshed is an alternative way to map and conceive of the city (see www.walkscore.com/walkable-neighborhoods.shtml). The World Bank has begun a pilot project to measure the walkability of cities in the developing world (www.cleanairnet.org/caiasia/1412/article-60499.html).

For too long, states have supported policies that promote car-dependent cities and suburbs, the construction of complex road projects, and land-use planning strategies where retail plazas and businesses are located far from residential areas, effectively discouraging the use of non-motorized transportation and walking. In many residential areas that have sidewalks, these are commonly incomplete and not meant to be used by pedestrians. In the

US vehicles are used overwhelmingly compared to other modes of transportation. Those who do not own a car must rely on either public transportation or non-motorized transportation to reach their places of employment, school, and grocery shop, and even for recreational purposes. Walking in such car-dominated cities has consequences. Cervero and Duncan (2003) found that the poor walk more compared to those with higher incomes, making them subject to higher risk for injuries. Brownson et al. (2001) also found that individuals from lower income groups walk more and in heavier traffic compared to those with higher incomes. Walking in the car-dominated city can be a dangerous enterprise. In the US, pedestrian fatalities are thirty-six times more likely than vehicle occupant fatalities per kilometer traveled. Pucher and Dijkstra (2003) show that walking and bicycling in the US is in fact more dangerous than in Europe, evidenced by the much larger mortality rates among pedestrians and cyclists compared to motor vehicle users. Many European countries show a more friendly approach to non-motorized transport users, integrating them into their transport systems.

The creation of a more pedestrian city is a difficult but not impossible goal. Engineering solutions, new planning methods, and changes in attitude are required. Above all, however, we need a re-examination of the city, less as vehicle-dominated system and more as place that is convivial to walkers, as well as drivers, children, and car owners, people as well as machines.

Conclusion

Across the globe there is an epidemic of traffic injuries and fatalities that is particularly marked in cities of low- and middle-income countries. Almost 90 percent of traffic fatalities occur in low- and middle-income countries. The most vulnerable in cities around the world are pedestrians, cyclists, and public transport users. The poor, the young, and the old are at especially high risk.

There is enough empirical evidence to reach firm conclusions to reduce traffic accidents for pedestrians. They include:

- limiting and controlling speed through active and passive measures that target drivers' decisions on how fast to drive, and incorporate road conditions that indirectly force drivers to reduce their average speeds;
- organizing traffic away from residential areas, limiting inner city and business area traffic flow, and encouraging alternative modes of transportation that focus more on health benefits and non-motorized road users;
- continuing public information programs that highlight the safety of high-risk groups and vulnerable urban populations;
- integrating urban planning and public health concerns to design built environments that promote healthier lifestyles rather than just safer behaviors.

Figure 6.4 Alternative urban transport: gondolas in Venice. Walking only cities
　　　　　 have a special charm.

Photo: John Rennie Short.

Pedestrian injuries are not accidents. They are preventable. If the vulnerability
of the human body was the determining factor, then it is unlikely we would
design our transport system the way we have. "Accidents" reflect and reinforce
social differences: they are less accidents and more manifestations of wider
and deeper inequalities in society that reflect the relative power of a vehicle-
dominated, as opposed to a pedestrian-dominated, culture. By reimagining
the city as a series of walksheds we refocus on the citizen as a walker and
the city as a place of walking (Figure 6.4). We need an anti-Marinetti to write
a program for the future in which the city reflects the needs of the pedestrians
and the frailties of the human body rather than the needs of the drivers and
the power of their vehicles. It is perhaps fitting then that the same country
that produced a modernist Marinetti is also home to the late modernist
CittaSlow movement.

CASE STUDY 6.1 **Dangerous by design**

In the first decade of the twenty-first-century vehicles killed almost 42,000 pedestrians in the US. Almost 4,000 of them were young people aged under sixteen. These were less "accidents" and more the result of streets and cities designed more for vehicles than for pedestrians. A pedestrian danger index (PDI) that measures the number of pedestrian deaths relative to the amount of pedestrian activity, as defined by those in the 2000 Census reporting that they walked to work, was calculated for different metro areas in the US. The most dangerous city in the period 2007–2008 was Orlando where there were almost 2.9 pedestrian fatalities per 100,000 residents. There was relatively little pedestrian activity in the city but a relatively high proportion of fatalities. Table 6.1 lists the ten most dangerous metro areas. They tend to be Sunbelt cities designed almost entirely around the needs of vehicles rather than pedestrians. There is also a compounding demographic reason for the top four cities being in Florida. Any ideas?

Across the country little more than 5 percent of federal transportation funds are used for ensuring safety of pedestrians and walkers. And yet in metro areas throughout the US between one-fifth and one-third of all traffic deaths were pedestrians.

The suburban shift of activities is constructed around arterial roads designed to move traffic through as quickly as possible. Often pedestrian walkways are small, inadequate, or sometimes completely lacking. The most dangerous metro areas are the new areas designed only with the needs of vehicles in mind. The pedestrian fatalities are less accidents and more the end result of an emphasis on designing cities and streets to ensure vehicle speed rather than pedestrian safety.

Table 6.1 Most dangerous metro areas (over 1 million residents) for pedestrians

Metropolitan area	Deaths per 100,000	PDI
Orlando-Kissimmee, FL	2.86	221.5
Tampa-St Petersburg, FL	3.52	205.5
Miami-Fort Lauderdale, FL	3.04	181.2
Jacksonville, FL	2.61	157.4
Memphis, TN-MS-AR	1.83	137.7
Raleigh-Cary, NC	2.02	128.6
Louisville, KY-IN	1.93	114.8
Houston, TX	1.81	112.4
Birmingham, AL	1.30	110.0
Atlanta, GA	1.37	108.3

Source: http://t4america.org/resources/dangerousbydesign/ Accessed 22 September 2010.

CASE STUDY 6.2 **Traffic accidents and civil society**

In rapidly modernizing cities traffic accidents reveal as much about the distribution of power and the health of civil society as they do about the state of urban traffic.

In February 2010 there was a car accident in Moscow. A Citroën and a Mercedes collided. Two women in the Citroën died while the occupants of the Mercedes, including the Vice-President of the country's largest oil company, escaped unhurt; nothing unusual in a city paralyzed by rapidly rising car owner-ship, a chaotic road network and poor driving behavior. In 1991 there were sixty cars per 1000 residents. By 2009 the figure had increased to 350. There was almost no planning for the increase and the road system, radial avenues eman-ating out from the Kremlin, exacerbated traffic jams. Driver behaviors, which in the West had taken decades to evolve, emerged in a country where the lack of a civil society was evident on the roads. Drivers routinely block intersections, park anywhere while police and state officials routinely break traffic regulations. The police initially reported the accident as the fault of the Citroën driver. It was later revealed that the Mercedes, with a siren blaring, had crossed into the central lane reserved for police and emergency vehicles. It was typical driving behavior for the wealthy and well-connected of Moscow who regard traffic regulation as something for only the poorer people to observe (Gessen 2010).

The growing disparities in rapidly growing and modernizing societies are visible on the streets. In China, as accidents between rich car owners and poorer bicyclists and pedestrians increase, so do the political implications. On June 26, 2005, Lui Liang, a student, was cycling in the Chinese city of Chizhou. A Toyota sedan, driven by a wealthy investor and well-connected administrator, Wu Jinxing, hit him. In the ensuing dispute the investors' two bodyguards, who were riding in the car, beat the young student. This minor incident sparked a large-scale urban protest. Wu and his bodyguards were taken to the police station. A large crowd formed outside station eager to see the rich outsiders punished. Almost 10,000 people milled around the station and riot police were called. When the crowd thought that the bodyguards were being treated too leniently the crowd threw rocks and bottles at the police. The crowd, now emboldened, overturned Wu's car as well as two police cars, set them alight and ransacked a supermarket. The rioting continued until the evening. At 11 p.m. 700 riot police arrived and the looting ended. A rare event to be sure, but in cities around the country, accidents between wealthy car owners and poor bicyclists and pedestrians take on a wider significance than just an "accident."

7 Modernity and urban utopias

The early Bauhaus

Direct lessons from the past are difficult to infer. In our era of rapid transformation, what happened in the past is rarely a guide to the present or a model for the future. The English novelist L. P. Hartley reminds us that, "The past is a foreign country; they do things differently there." The further back in the past, the more foreign. However, there are echoes and resonances that can inform our thoughts and influence our practices. Walt Whitman once remarked that "History does not repeat itself, but it does rhyme."

Modernity typically retreats from the most recent past while at the same time constructing a historical consciousness out of more distant memories. What lessons can we draw upon from this formulation? As we seek not only to make sense of the modern city but also to build a better contemporary city, a useful place to start is the Bauhaus. The recent celebrations of the art school's ninetieth anniversary focus attention on its legacy. The Bauhaus's fragile tenure was equal to the Weimar Republic's short duration, from 1919 to 1933. In a contemporary world of unceasing change, an obvious question to ask is: does the Bauhaus have anything to teach us about contemporary urbanism and in particular, given its utopian credentials, the possibility of a more humane city? Anything? Anything at all? What are the rhymes of the Bauhaus that make sense for us today?

It began with a call to collaborative arms. In the 1919 "Bauhaus Manifesto and Program" that inaugurated the Weimar Bauhaus, the school's Founder and first Director, Walter Gropius, likened the new school's mission to that of medieval cathedral building:

> Together let us desire, conceive, and create the new structure of the future, which will embrace architecture and sculpture and painting in one unity and which will one day rise towards heaven from the hands of a million workers like the crystal symbol of new faith.
>
> (quoted in Bergdoll and Dickerman 2009, p. 64)

The Manifesto and Program's cover illustration is Lyonel Feininger's woodcut, *Cathedral*, its elongated and intersecting planes punctuated by sparkling stars, refined versions of the cruder, glass-shard surfaces of German Expressionist woodcuts. Feininger's woodcut infused the Bauhaus at its very beginning with the German modernist tradition of signifying crystallization. The angular crystalline grid appears and reappears in many forms in the work of the early Bauhaus artists, especially that of Paul Klee, who often applied prismatic, interpenetrating planes to the representation of his playful versions of urban growth, as in his 1921 *Dream City*.

The Gropius quote summarizes the hopes of the early Bauhaus for the creation of a new society through, by, and in new design and architecture. Shorn of the ornamentation and class codings of the neoclassical, these modern buildings were to constitute the new city to house the new society and embody a utopian future.

Early origins

The Bauhaus grew out of crisis: it was in part a response to the new machine age that emerged from the horrors of World War I and to the political rupture of the German and Russian revolutions. The Bauhaus was born out of the trauma of loss and defeat, unceasing change and uncertain futures. The instability of the times can be measured by the hyperinflation of Germany's currency. When the school opened in 1919, the US dollar was worth 7.9 marks; in early 1923 a dollar was worth 7,260 marks and by the end of the year it was worth 4.2 trillion marks.

One standard interpretation of the Bauhaus is as a crucial ingredient in the development of twentieth-century international modernism, a synonym for decontextualized modernity, that unfolds in particular through high modern architecture's inauguration of clean-cut lines and its dismissal of ornamentation. This view of the Bauhaus equates it with the International Style. No more bourgeois pediments—this Bauhaus dictates the rational city of tall, flat-topped buildings laid out to maximize mobility. Bauhaus aesthetics structure the city as abstract rational form, built from scratch and erasing the demolished past. This is the tale told by Tom Wolfe in his 1981 entertaining book *From Bauhaus to Our House*: Bauhaus as the incubator of the International Style with Walter Gropius and Mies van der Rohe moving on from Germany to the US in order to influence the wider world down to the present day. The International Style, while it produced some beautiful buildings, failed to create livable cities: its downfall was in the top-down designs by authoritarian architects shaping cities that people must, but could not, live in. The inherent bankruptcy of this wing of modernist architecture and city planning now makes this putative Bauhaus legacy easy to criticize. Gropius' statement on the construction of a crystal city is thus reread as the blueprint for a top-down imposition of a rigid design. In this Bauhaus-as-International-Style narrative, the flawed early plans and muddy ideals lead invariably to the second-rate modernism that most city dwellers have to live with, the failed housing projects, the crappy town centers, the inhumane cities, cheerless places, and sad spaces. The conclusion leads to the rejection of modern architecture, modernist urbanism, and the legacy of the Bauhaus. In a revisionist retelling the heroes are now those who were early resisters to the Bauhaus. From 1937 to 1952, Gropius was an important figure in the Harvard Graduate School of Design, influencing a generation of urban planners and designers with his modernist beliefs. Pearlman (2000) compares the works and writings of Gropius with Joseph Hudnut, the Dean of the School, who at first encouraged Gropius to come to Harvard but then parted company with the German émigré. Hudnut in Pearlman's telling is now the hero for invoking the humanistic traditions of civic design, predating and prefiguring a more late modern/postmodern urban sensibility against the rationalist modernist Gropius whose ideas dominated postwar US architecture, planning, and urban design. Hudnut was defeated in the short term, but was validated in the long term. History, in this now dominant narrative vindicates the early postmodern and trashes the modern. Each country will have their own modernist architect and postmodern urbanist as respective folk devil and local hero. In the case of the US, it involves the rejection of Robert Moses and the embrace of Jane Jacobs (Short 2006).

The lessons from the Bauhaus are thus to avoid the simpleminded views of buildings and cities and to reimagine the complex relations that link people with urban space. The lesson of the Bauhaus, then, is as a failure, a creative failure that provoked a more humane postmodern architecture and urbanism in the reaction and resistance to its ethos. This lesson is an important part of the legacy of Bauhaus, but not the only one.

The telling of the Tom Wolfe morality tale is, of course, an oversimplification. It is the dominant narrative from a European or North American perspective where modernist architecture very early on reached dominance and then quickly succumbed to intellectual and aesthetic bankruptcy. Anthony King (2004), with his careful reading of non-Western urbanization, however, reminds us of the multiplicity of urban modernities around the world. Modernism was inextricably linked to connections with the non-West. Moreover, there are multiple modernities—hybrid, vernacular, and gendered. And the modernist city survives in the search for world city status that provokes the feverish new building in Dubai, Seoul, and Shanghai. The Bauhaus project also lives on in the more complex reading of multiple modernities. Take the case of Tel Aviv where 2,000 modernist structures, one of the largest single collections of the early International Style of the Bauhaus, were declared a UNESCO World Heritage Site. The story is convoluted, as befits a more subtle appreciation of the different folds of Bauhaus legacy, as the work of German architects was built, then ignored, and then preserved in the Jewish state, with the subsequent UNESCO designation full of contested meanings for Jews and Arabs (Rotbard 2005).

A more complex picture

I want to add another, more complex picture to this simple tale of Bauhaus as an early progressive architectural force that quickly solidified into an authoritarian imposition of an international aesthetic shorn of local connections and national traditions. I want to excavate and highlight the story of the early Bauhaus, especially the Weimar Bauhaus. I will argue that here was a more sophisticated, somewhat whimsical, complex understanding and representation of the city that has a greater contemporary relevance for us today. The early Bauhaus provides a model for the Habermasian project of redeeming modernity.

The Bauhaus was neither static nor fixed. From its foundation in 1919 to 1925, the school was based in Weimar; then it moved to Dessau and then to Berlin in 1932, where it lasted only briefly before being closed by the Nazis in 1933. It had a variety of Directors, including Walter Gropius, Hannes Meyer, and Mies van der Rohe. The guiding philosophy also changed from a Neo-Romantic to a more functionalist notion and from an opposition to an embrace of commodity.

The Bauhaus is neither easy to summarize nor simple in its message. However, we can make a distinction between the early Bauhaus of Weimar and the later Bauhaus of Dessau and Berlin. From 1919 to 1925 it was based in the provincial city of Weimar, the newly designated national capital. Here Gropius assembles a large range of characters from the European avant-garde that includes Johannes Itten, George Muche, and Oskar Schlemmer as well as Paul Klee and Wassily Kandinsky. This Bauhaus, much closer to the appalling experience of World War I, was still finding a shape and form; it was a place where there was more experimentation than the later Bauhaus. It was freer, sillier; it flirted with Dada. It was the Bauhaus of color, mysticism, humanism (especially in terms of the human body as spatial reference point). This Bauhaus was more in opposition to the market and the state and had a greater distrust of the machine and the commodity than the later Bauhaus.

In Weimar, especially until 1923, the Bauhaus encouraged the mystical as well as the rational, a concern with the human body as well as the abstract form. There was the influence of Johannes Itten, for example, a Swiss artist appointed Bauhaus Master of Form who was part charlatan, part mystic, dressed as a monk in clothes that he designed for himself, followed unusual dietary practices including fasting, and encouraged students to connect with inner creativities through breathing and play and meditation. Itten was not alone in his eccentricities. Others included Gustav Nagel who wanted to ban upper case lettering, and Rudolf Hausser who persuaded some of his students to join him on a wandering pilgrimage. Then there was Oskar Schlemmer who came to Weimar in 1920 and produced balletic works as well as elaborately sculptural costumes for the dancers. His work was concerned with the spatial expression of the human body. Another early Master of Form at Weimar was Georg Muche who worked in the Expressionist style but also had mystical leanings. He designed the Haus am Horn—a cubic construct with fittings designed by Moholy-Nagy and Marcel Breuer. It was the prototype of cheap mass-produced housing but with exacting design standards and an emphasis on comfort, economy, craftsmanship, and design (see Figure 7.1).

Figure 7.1 Early modernism: Haus am Horn, Weimar.
Photo: John Rennie Short.

One of Itten's students, Paul Citroen, produced a collage in 1921 entitled *The Big City*. The multiplicity of angles and lines evokes a noisy cacophony of a city, a place of myriad viewpoints and experiences. This is not the rational city of the building as machine or the city as plan, but the city as a metropolitan diversity of shapes and forms. Even the later Bauhaus carried on with some of these themes. The distinction between the early and the later Bauhaus is more of a shift than a sharp break. Marianne Brandt's 1926 photo-montage, *Our Unnerving City*, also depicts a city of overlapping images and urban experiences. It is perhaps too much of a stretch to see an explicit post-modern sensibility in the early Bauhaus but the movement was still so full of promise, so unanchored into a single vision, still so moving through an uncharted rather than a carefully mapped artistic landscape, that more things seemed possible than the simple message of the later Bauhaus.

The early Bauhaus has a wider angle of vision on the city and the urban experience than the later Bauhaus of the International Style. The diversity arose from the self-conscious transdisciplinary nature of collaboration in the early Bauhaus. Architecture was only officially taught beginning in 1927. Before that the city was subject to a wider aesthetic than the plans of architects or town planners. The early Bauhaus in Weimar was marked by its inter-nationalism and interdisciplinarity. The teachers were drafted from around Europe and included Paul Klee, a Swiss painter of German nationality, Lyonel Feininger, born to German parents but who grew up in New York, Itten who was Swiss, the Moscow-born Wassily Kandinsky, and the Hungarian Laszlo Moholy-Nagy. In the Weimar Bauhaus, all students began with Itten's notorious *Vorkurs* (preliminary course), which emphasized principles of form, color theory, and the characteristics of materials. There were workshops on weaving, cabinet making, pottery, typography, and sculpture, among many others. Paul Klee taught bookbinding. It was only in the later Bauhaus of Dessau and Berlin that there were any formal classes on architecture. This representative sampling of the European avant-garde was a transdisciplinary, international group. And while deeply gendered and racialized, its members provided at least an indication of what a more cosmopolitan embrace of urbanism would look like. Therein lies some of the ongoing fascination with the Bauhaus. Although short-lived, it provided a tantalizing reminder of what a community of aesthetic intellectuals can produce.

There was an ingrained humanist urbanism in the early Bauhaus that provides us with a subtle sense of how the human is imbricated in a kind of playful, creative community rather than with the notion of the building blocks of the International Style. The Gropius quote suggested that the Weimar Bauhaus was not only a community of artists but also a collective endeavor that referred back to the tradition of cathedral building that included architects, designers, painters, sculptors, and craftspeople with a devotion to cathedral building as a collective endeavor, with a deep understanding of each of their arts as part of a whole. Urban communities were also envisaged in the plan for a Bauhaus housing settlement in Weimar. In 1920, one of the students, Walter Determann, presented this vision in a plan that incorporated the settlement into the existing urban structure, anchoring the buildings with community spaces. Determann's highly stylized design for this housing scheme almost evokes theosophical charts and diagrams as well as presages the Garden City aesthetic: its sheer beauty is a fine example of the Bauhaus' dedication to a unity of art and design. Determann's richly textured designs were never implemented. Instead, only one house was built for the "Bauhaus 1923" exhibition, the Haus am Horn. Thus, the realized legacy is but one modernist building standing in for the complete community.

The early Bauhaus also had more explicit connections with the past than its later models maintained. The standard view interprets the whole Bauhaus according to the standards of the later Bauhaus in terms of a decisive break with the past. In fact, the early Bauhaus looked farther back, past the horrors of the recent past to an idealized, prewar, premodern, pre-industrial society for inspiration. In 1921 Marcel Breuer, best known for his chairs of modular metal and fabric, collaborated with future Bauhaus Weaving Master Gunta Stölzl and designed a chair that drew upon Hungarian peasant designs for wooden furniture. While later Breuer chairs have become iconic examples of modernist furniture, this chair from the early Bauhaus is an homage to premodern designs and craftsmanship. It is called the African or Romantic chair.

The early Bauhaus also had a wider worldview than the rationalist modernist belief system of the later Bauhaus. Itten, who was part mystic, encouraged his students to take physical exercise before artistic practice. Kandinsky held theosophical beliefs and his work is inspired by the transcendental quality of form and color. No mere technician of the abstract, Kandisky's work was embedded in a deeper belief system. Colors were not just different parts of the spectrum: they were linked to emotions and feelings.

The human form also figures more largely in the early Bauhaus. Puppets were made. Oskar Schlemmer created the *Triadic Ballet* in 1924, where figures dressed in geometricizing costumes dance in a grid. Kandinsky drew designs for ballet movements in which the human body appears as a series of vibrant lines indicating movement more than form. Klee's whimsical figures inhabit magical spaces.

And there was a joyful, playful quality to the early Bauhaus. This is evident in the parties and social gatherings that so shocked the solid bourgeoisie of Weimar as well as in the objects produced, including puppets, chess sets, and children's toys. Children's toys with geometric forms and bright primary colors were displayed in the Haus am Horn. The early Bauhaus envisaged a ludic city. Even the stark geometry of the grid was softened and colorized by Kandinsky and Klee. There is a more whimsical attitude to geometry and form than the more austere architectural isometric lines of Gropius and Rohe. The Bauhaus as International Style lacks not only a sense of humor but also a basic sense of joy. The early Bauhaus was full of caprice—even silliness—sometimes juvenile but always entertaining and joyful.

And then there is the issue of gender. Half of the students at the Bauhaus were women. Women were particularly important in the weaving workshop, although Alma Buscher's wooden toys and Marianne Brandt's photomontages indicate that women were not restricted to textiles. Gunta Stölzl, the only female Master, was in charge of the weaving workshop. Yet Bauhaus textiles, especially the colorful, woven grids of Anni Albers, are often overlooked in the Bauhaus-as-International Style narrative. It is the male visionary architect who appears as dominant, yet the female weaver was also experimenting with form and color.

My basic argument is that today's dominant narrative of the legacy of the Bauhaus only draws upon a partial and very limited narrow range of the Bauhaus experience. It foregrounds the later Bauhaus, it concentrates on architecture and downplays issues of community, play, spiritual and philosophical complexity. It ignores the embrace of the past in the early Bauhaus, its ludic quality and sensibility, its fascination with the human form. This is no accident. The Bauhaus as International Style was solidified in the US, by Moholy-Nagy and Mies van der Rohe in Chicago and Gropius in Harvard. A different Bauhaus would be remembered if Itten or Schlemmer or Muche had created and transmitted their Bauhaus myth.

Rhymes and resonances

It is important to recall this other Bauhaus, this Bauhaus of color, community, concern with the human body, mysticism, and downright wackiness. There was a romanticism and mysticism at the heart of Weimar Bauhaus. And it is this more complex Bauhaus that I think creates resonances with the present day. This more nuanced Bauhaus is more sensitive to the past, the natural world, and the mystical world than the isometric architectural drawings of Gropius and the clean lines of "less is more" Mies van der Rohe.

But what does this more complex Bauhaus have to teach us today? Less by way of direct lessons, more in the form of suggestive resonances, interesting echoes. They defined themselves and indeed were part of an international avant-garde who wanted to transcend (and oppose) narrow nationalism. Today the intellectual's distrust of globalization has led to a concern with the local and the vernacular. But the Bauhaus reminds us of the liberating and oppositional force of a cosmopolitan, intellectual transnationalism.

The Bauhaus also reminds us of the benefits of interdisciplinarity. The problems of the International Style were caused largely by the separation of urban design from the varied discourses of the early Bauhaus. It was an aesthetic and practice stripped of its color, sense of community, and love of play. The designs of buildings and cities were removed from their context and situated in a narrow technical discourse of the expert, the architect as visionary, the planner as god. A more humane city will only arise from a wide-ranging conversation among the many rather than from a technical manifesto created by the few.

The earlier, Weimar Bauhaus had a stronger sense of the organic than did the machine aesthetic of the Dessau Bauhaus. We can reclaim that general concern by looking once more at the connections between cities and the environment. A sustainable urban future requires a greater concern with the natural world, always bearing in mind that nature is not all that natural and the city is anything but unnatural, a notion that the crystal figures so very well.

Crystal city, liquid city

There is one area, however, where the Bauhaus' early and later versions have less relevance. The idea of the crystal city was a powerful one. The image of the glass crystal appears in much utopian thought in the early twentieth century part of the many urban utopias expressed in specific spatial forms from Tatlin's 1920 *Monument to the Third International* to Sant'Elia's ideal Futurist city and Bruno Taut's utopian designs. Glass featured largely in Taut's designs; it had mystical and transcendental qualities. The crystal, a natural form that was also geometric, seemed to combine the organic and the rational in one beautiful form. But there is a problem at the heart of this spatial form as type of utopia, the creating of a better society through the construction of a fixed form. In fact there are two problems.

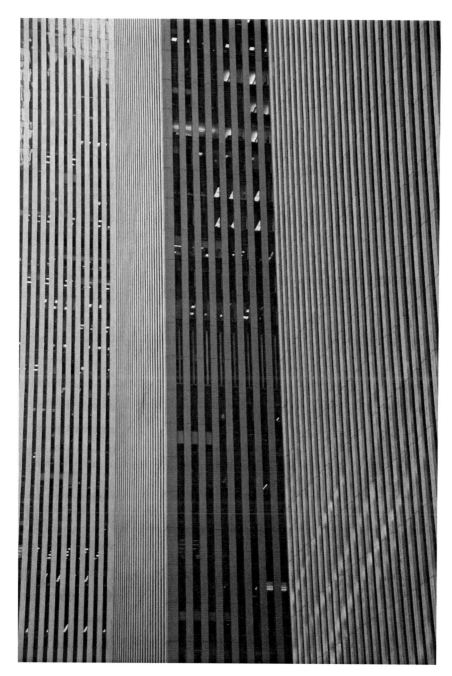

Figure 7.2 High modernism: Sixth Avenue, New York City.
Photo: John Rennie Short.

The first lies in the nature of contemporary urbanism. And here I need to introduce the counter-notion of the liquid city to the crystal city of the Bauhaus. I draw upon the observations of the sociologist Zygmunt Bauman, whose book *Liquid Life* (2005) describes the precarious life lived under conditions of constant uncertainty. Bauman uses the term to refer to time and the question of identity in a rapidly changing world, but I also think it is appropriate to use with reference to space and the incoherence of the sprawling built environment. Metropolitan growth, especially now, has a liquid quality. To be sure, large capital investments can anchor the urban form for some time. Large fixed investments such as airports and major downtown redevelopments, for example, can "fix" the city for a while, influencing future cycles of investment. But the amount and speed of capital flows, the global urban investment opportunities and the consequent rapidity of changing surfaces of profitability, at all scales from the global, national to local, can soon turn the fixed into the irrelevant, the profitable into the unprofitable. Urban fixity and metropolitan flow are the elemental forces shaping the contemporary city. The future is constantly being created and continually recreated as capital investments and disinvestments flow over the built form, at times reinforcing the existing form, at others reconfiguring the individual city as well as whole networks of cities. Cities arise from the desert, as others lose population, old urban industrial regions wane as others arise in the Far East, and new centers of finance flourish while others languish (Figure 7.2). To be sure, there are sites of fixity. London and New York, for example, continue to dominate the global urban hierarchy of advanced producer services, while up-market neighborhoods in cities around the world often maintain and sometimes reinforce their exclusivity. Our urban world is composed of sites of fixity in a liquid flow of change. The Bauhaus, in contrast, was conceived in a world where urban fixity dominated metropolitan flow. We live more in the space of flows. We live in an era of rapid urban reconfigurations as individual cities can rapidly move up or down the national and global hierarchy and individual parts of cities can be both revalorized and devalorized. In Chapter 11, as an example, I describe Megalopolis, the urbanized northeast corridor of the US as a large liquid metropolis whose boundary demarcation is always provisional. I have also used the crystal/liquid city binary in Table 7.1 to organize some of the differences between modern and postmodern urbanism.

And here we encounter the second problem with the crystal city. The Bauhaus, like many utopian ideas and practices of the last two hundred years, supposes the ideal city as a fixed destination not a journey, a hard structure not a liquid process, a crystal not a flow. But if the better life is only imagined through a fixed form then the spatial form of urban utopia needs to be created and maintained, and here it can easily slide into authoritarianism. An alternative model sees utopia as a journey not a destination, a process not a fixed form. Theorizing the city as a place where ordinary people can lead dignified and creative lives (Short 1989) is to generate a more open and democratic

Table 7.1 Crystal city/liquid city

Crystal city	Liquid city
Utopian	Non-utopian/disutopian
Fixed	Loose
Clear	Cloudy
Future	Present/past
Hope	Anxiety
Class	Market
International	Global
Berlin, Brasilia, New York	Dubai, Las Vegas, Shanghai

"utopian" rather than its social engineering connotation of achieving and maintaining a fixed form.

Is this it?

The Bauhaus continues to have relevance because it asked that perennial question. Is this it? Is this, as a society, the best we can do? Are our cities the best possible cities we can build or even imagine? The question is not restricted to the Bauhaus. It is the question that prompts religions and social movements as well as utopian visions. But it is the question that each generation needs to ask itself. Because without that question we will continue to live in the cities made by and only for the rich and the powerful. The recurring question is always posed in a particular context. The Bauhaus was born out of the traumatic brutality and sheer horror of World War I, where casualties amounted to thirty-seven million—sixteen million dead, and twenty-one million wounded. Germany had 2.4 million dead and 4.2 million wounded out of a population of sixty-five million. The total German casualties amounted to almost 10 percent of the entire population. A similar percentage figure for the contemporary US would give an absolute number of almost thirty million. Everyone would have been affected, directly or indirectly by death, loss, suffering, dismemberment, and grief. The scale of the savagery is almost too much to contemplate and that is what the Bauhaus did. They wanted nothing to do with the recent past. The recent past was death and suffering, only the look into the future provided a better prospect; the past was too terrible to contemplate, it needed to be distanced and forgotten. The architecture had such a historical amnesia because it was the only way to imagine a better world. Gropius, the German cavalry officer, never spoke of his war experiences on the front. He committed the rest of his life to erasing the past, forgetting the past in a constant homage to the future.

Our context is the opposite. We, in the West in particular, are faced with a period of relative stability and peace. It is the future that is our worry. Images

Figure 7.3 Late modernism: contemporary house in Weimar.
Photo: John Rennie Short.

of the apocalypse dominate popular culture and contemporary imagination. We are so anxious about the present and future that we idolize the past. We recycle, reuse, reappropriate. We eschew the future and the difficult job of imagining a better future. Postmodern architecture and postmodern urbanism are in their different ways responses to the anxiety and fear of the future (Figure 7.3).

There is renewed interest in utopian thought. David Pinder (2002), while admitting the authoritarian nature of much urban utopianism as well as the retreat from utopian thinking, reminds us of the need to open up the debate on more imaginative urban utopias. And here the Bauhaus has something to teach us. They, despite their recent historical experience, had the optimism and courage to imagine a better future. They wanted to build the "cathedral of socialism," the original name for Feininger's woodcut. That their future was flawed and disappointing is, for us, now irrelevant. They made their mistakes as we will make ours. But they employed the historical consciousness that infuses modernism with memories appropriate for constructing a utopian future, lifted their eyes from the present, and saw hope and life and the possibilities of a better future.

Even in the later Bauhaus, the formal architectural classes of Hannes Meyer "stressed humble materials over sleek forms and prioritized small clusters of human activity over the large geometric compositions favored by his predecessor" (Ouroussof 2009). Meyer's socialist leanings ensured his dismissal by the Nazis.

The Bauhaus, like us today, stood at the edge of change. As Gropius wrote in 1919, "The old forms are in ruins . . . We float in space and cannot perceive the new order" (quoted in Wilk 2006, p. 11). If we lift our eyes to imagine a more just and verdant city, not a crystal city of fixed dimensions but a liquid city of possibilities and contingencies and surprises, then we will have rhymed with the early Bauhaus in carrying on the creative tradition of imagining a more humane city.

CASE STUDY 7.1 **Weimar, East and West**

Some dates resonate so deeply that the numbers alone suffice. Nine eleven. The reversed numbers, eleven nine, have less resonance but the date marks an equally important global event. On November 9 in 1989 the Cold War began its final decline with the toppling of the Berlin Wall.

Twenty years later to the day, I crossed the old border between East and West when I left Weimar by train very early in the morning to arrive at Frankfurt. I arrived the week before for a conference in Weimar, a city imprisoned on the east side of the old border. The city is one of the historical centers of German culture. The home to Cranach, Goethe, Schiller, and Herder. The town is filled with memories of classic German culture as well as fragments of the modern movement. Goethe's house is a site of pilgrimage. The historic Haus am Horn built in 1923 stands as one of the very earliest examples of modernist domestic architecture. The modern soon turned into the dark and savage. From the Bauhaus studios where Kandinsky, Klee, and Moholy-Nagy taught and argued you can see the tall grey memorial tower on the site of Buchenwald concentration camp.

Leading up to 11/9 in 2009, the newspapers and media were filled with images of East Germans streaming across in their thousands, people partying and spontaneously chipping away with hammers and picks at the symbol of oppression. The Wall has long gone but the integration between East and West is still a process. Unemployment is still twice as high in the East as in the West. A solidarity tax is still leveled on all citizens to bring the East up to the standard of the West. The legacy of disparity in investment is evident in the newer roofs in the villages and towns of the old West compared to the old East with its abandoned factories and depopulating towns. But there are also signs of a smoothing out. Weimar boasts lots of new buildings and the whole historic center was spiffed up for its designation in 1999 as European Capital of Culture. With more rich elderly West Germans moving into the city, it is becoming an affluent retirement town.

I spoke with some people who had lived in the Weimar of East Germany. They remember the polluted smell of coal-fired power stations, the lack of goods and the restricted freedoms. But the memories are disappearing from view. Ex-Stasi (secret political police) are now running for office, asking voters to put the past behind them. Early in the morning of 11/9 in 2009, as I left the hotel to start my long journey home, I asked my taxi driver if he could remember the great day twenty years ago. No, he said, I was only five. A new generation is coming of age that only knows one Germany. As the momentous day fades into historical memory we will forget the seeming solidity of what was before.

———————————

8 Big city

In Chapter 3 I considered various permutations on the connections between city, country, and world. Here I look at the relationship between city and country, and focus on one permutation: *Big City/Small Country*. To be more precise: there are some countries where one city dominates such that it contains most of the population and is the sole center of power and influence; this state of affairs is termed *urban primacy*.

Consider the case of Haiti, where, before the catastrophic earthquake in 2010, the population was almost ten million. Almost one-third lived in the capital city of Port-au-Prince. When the earthquake hit the city in March 2010 it paralyzed the whole country since much of the administration centered in the capital. It was not just one city that was impacted but, because of the urban primacy, an entire country was devastated.

In a classic paper Mark Jefferson (1939) identified primate cities. He defined them as cities of a different order of magnitude and significance from all the other cities in a national urban system. He gave the examples of Vienna, which had twelve times the (1934) population of the second city in Austria; Copenhagen, which had nine times the (1935) population of the second city in Denmark; and London, which had a (1931) population seven times larger than the second city in Britain. Later papers sought to identify the conditions under which such primacy occurred. Mehta (1964) found that primacy was a function of the small area and population size of countries. Linsky (1965) proposed that high urban primacy is most often associated with countries of small extent but high density, low per capita income, export-oriented and agricultural economies, a colonial history, and rapid rates of population growth. Ades and Glaeser (1995) correlated urban primacy with political instability and high levels of centralized political power. Moomaw and Alwosabi (2004), restricting their analysis to Asia and the Americas, showed that primacy is associated with economic and population size, economic development, population density, level of industrialization, and capital city status. Krugman's (1996) general model suggested that primacy decreases with the openness of a national economy, but this has not been supported in the literature.

The political importance of urban primates is also thought to influence economic growth and development. The desire by economic actors to be close to political leaders and administrators, termed "productivity effects" and the state's need to buy off the urban masses with public spending, "rent seeking," can skew economic development to the primate city.

Measuring primacy

In this chapter I want to reconsider urban primacy. The first step is to measure it. Easier said than done. The "dirty little secret" of global urban research is that the basic comparative data are not very good (Short et al. 1996). We have yet to establish a solid global metric of comparative urban data. The simplest measure, with the most accessible, most easily assembled data set with the widest global coverage, is still population size. Despite all its limitations, and these will be discussed, urban population size provides us with the—relatively speaking—most accessible, globally comparative urban data source. But it is important to bear in mind that it is not so much the best source as the least worst.

An obvious source of city population data is the United Nations (UN). The UN Department of Economic and Social Affairs, Demographic and Social Statistics lists the population of the principal cities for most countries in the world. This source has the benefits of ready accessibility, constant updates, and ease of use. The downside is that, as with all international urban data, it is deeply flawed. Different countries take censuses at different times, use different measures (e.g., some use formal urban political boundaries while others give the data for functional urban agglomerations), and have wildly varying degrees of accuracy. Take the case of Nigeria and its major city of Lagos. At the UN source, accessed in February 2007, the city's population is available for only 1975 and is listed at just over one million. According to the latest official figures in Nigeria the population of Lagos in 2006 was almost nine million, but local officials accuse the federal government of deliberately undercounting this southern Nigerian city and its encompassing region to favor the northern part of the country (Sanni 2007). More neutral observers estimate the current population of Lagos between thirteen and fifteen million (Packer 2006). The UN source also only lists cities of more than 100,000 making it difficult to measure primacy in countries with relatively small cities.

A more user-friendly source is available at www.citypopulation.de/ which lists the population of all principal cities for most countries. The data are drawn primarily from official censuses and estimates but even the website reminds the user that the data are all of varying, and some of suspect accuracy. This is the main data source used in this chapter.

Another major problem with urban population data is the distinction between population estimates for the city and for the wider agglomeration of the functioning metropolitan region. Because all countries have data for the cities but there is only limited coverage of metro areas, this chapter will rely on city data. Primacy is often undercounted using just city data.

The "dirty little secret" of global urban research persists. Population censuses are of varying degrees of accuracy, are taken at different times, and have differing definitions of city and metropolitan regions. The figures that follow in the rest of the paper are more estimates than precise figures, averages across a wide band of error, and more appropriately used as rough comparisons rather than precise absolute values. The partial and provisional nature of the results should be in the reader's mind throughout the rest of this chapter.

There are a variety of measures of urban primacy. Jefferson (1939) used the ratio of the largest city to both the second and third largest city, Linsky (1965) used the ratio of the largest to the second largest city, while Mehta (1964) used the percentage of the population of the four largest cities residing in the largest city. We calculated two measures of urban primacy, the ratio of the population of the largest city in a country divided by the population of the next two largest cities. And the ratio of the population of the largest city in a country divided by the population of the next four largest cities. Since there was little difference between the two—more precise statistical details are available in Short and Peralta (2009)—we will only use the first measure.

Table 8.1 notes all the countries with a primacy value above 3.0. We have only included countries with a population of at least one million and thus have excluded Samoa as well as Antigua and Barbuda because their respective populations were only 172,000 and 74,000. A value of 3.0 and above tells us that the primate city's population is more than three times the combined population of the next two largest cities. This is a substantial concentration. The high threshold was used to offset the data problems already discussed. The value is high enough so that data issues do not lead to the inclusion of non or only marginally primate distributions. The precise figures are probably underestimates of the degree of urban primacy, since we used city rather than metropolitan data. The figures are best used in a comparative, suggestive discussion rather than as precise, accurate measures.

The most extreme case of primacy is found in Thailand where the primate and capital city of Bangkok holds almost ten times the combined population of the next two largest cities.

Table 8.1 also records each country's status on the World Bank fourfold economic classification of national economies based on income per capita into low income ($905 and less), lower-middle income ($906–$3,595), upper-middle income ($3,596–$11,115) and high income ($11,116 and above). The classification is crude but provides a quick and useful indicator. Out of forty-one countries with marked degrees of primacy, 49 percent are classified as low-income, 44 percent medium-income and only 7 percent high-income. Table 8.1 includes a range of countries at varying degrees of development, as defined by the World Bank, including low-income (Togo, Nigeria, Uganda and Ethiopia), as well as lower-middle-income (Peru), upper-middle-income (Argentina) and high-income (Austria, Greece, UK).

Table 8.1 Urban primacy and World Bank economic classification

Country	Primacy	World Bank classification
Thailand	9.48	Lower middle
Suriname	8.24	Lower middle
Togo	7.92	Low
Uruguay	7.37	Upper middle
Chile	5.98	Upper middle
Nigeria	5.94	Low
Uganda	5.94	Low
Ethiopia	5.82	Low
Mongolia	5.67	Low
Peru	5.43	Lower middle
Guinea	5.27	Low
Eritrea	4.98	Low
Namibia	4.80	Lower middle
Argentina	4.76	Upper middle
Mauritania	4.60	Low
UK	4.48	High
Hungary	4.47	Upper middle
Armenia	4.33	Lower middle
Côte d'Ivoire	4.24	Low
Madagascar	4.14	Low
Mali	4.04	Low
Nicaragua	3.87	Lower middle
Afghanistan	3.76	Low
Latvia	3.70	Upper middle
Georgia	3.49	Lower middle
Iraq	3.34	Lower middle
Myanmar	3.34	Low
Gambia	3.30	Low
Sierra Leone	3.30	Low
Austria	3.27	High
Haiti	3.26	Low
Greece	3.23	High
Jamaica	3.22	Lower middle
Tanzania	3.19	Low
Gabon	3.16	Low
Chad	3.14	Low
Central African Republic	3.10	Low
Costa Rica	3.09	Upper middle
Serbia	3.07	Lower middle
Cuba	3.04	Lower middle
Romania	3.02	Upper middle

Source: primacy calculated from data at www.citypopulation.de.

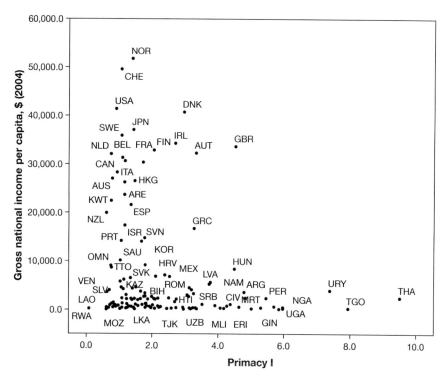

Figure 8.1 Primacy and national income.

Figure 8.1 plots primacy values against a more precise measure of national economic activity, gross national income (GNI) per capita, that is also available from the World Bank. There is no simple linear relationship between GNI per capita and levels of primacy. The overall relationship is more of a concave logistic curve. There are some broad conclusions to be drawn that undermine some of the more popular conceptions of urban primacy. Low-income countries exhibit both high and low values of primacy. High primacy is associated with countries with colonial as well as non-colonial histories; in large as well as small countries; in rich as well as poor countries; in politically unstable as well as politically stable countries.

The complex and non-linear relationships suggest that it is difficult to make conclusions using the full set of countries and looking for obvious empirical generalizations. This calls into question the current emphasis on statistical modeling using global primacy data sets. The paradox is that the more robust statistical analysis necessary for ever more sophisticated multivariate analysis is rendered both difficult and suspect by the quality of the data: specifically, measurement errors in the dependent primacy variable leads to a concern with the precision and statistical significance of slope coefficients. But more gen-erally, although regressions of the determinants of primacy do reasonably well, with R-squares of over 0.5, we suggest that these results collapse the different structural properties of primate countries in different stages of development. We suggest that such aggregate statistical analysis leads to a conceptual muddle about the determinants of primacy. We suggest it is more instructive to identify certain exemplar types. We will begin with the most pronounced case of primacy and then go on to consider a range of regional types including Latin primates, African primates, new primates, European primates, and non-primates. These are presented less as isolated case studies and more as building blocks in an early attempt at a more coherent theory of urban primacy. The exemplars are used as starting points rather than explanatory destinations.

Hyperprimacy

Among the hyperprimates, which we will define as those countries with a primacy value of over 5, there are some small countries such as Togo, but also large countries such as Nigeria. There are former colonies such as Uganda as well as the uncolonized Thailand. Size and colonial history in themselves cannot then explain urban primacy. Let us consider the most primate urban system to tease out some of the possible causal connections.

The population of Thailand exhibits the most pronounced degree of urban primacy. In 2007 it is estimated that out of a population of almost sixty-three million, just over fifteen million lived in the urban region of Bangkok. The city population of Bangkok was listed at 9.1 million; the next two largest cities were Chon Buri and Chiang Mai, listed at under 200,000 in each city. Thailand is one of the few countries in Southeast Asia that has not been formally colonized, so the colonial legacy argument is difficult to sustain. London (1979), however, suggests that a model of internal colonialism, with due regard to the exploitative policymaking of political elites, is more appropriate to explain the marked primacy. In other words there were powerful domestic political forces at work to shape Bangkok's dominance even before globalization added an extra push.

Bangkok's primacy accreted around a traditional center reinforced by subsequent elites, which in turn generated the basis and pathway for subsequent growth. Since 1782 the city has been the capital of the country. Bangkok was and remains the center of the country, the unchallenged center of political, intellectual, and religious life. The court is based in the city. There are few other towns of any size, in a predominantly homogeneous society. There is only one center of political power and permanent royal presence. In the past fifty years this embedded centrality of the city to the wider life of the country has shaped subsequent growth, especially export-led growth (Mutebi 2004). The city is the main transmission hub of economic globalization for the country and the wider region. The centralizing forces of globalization reinforced the national primacy. By 2005 the city was responsible for almost half of all gross domestic product. Thailand is one extended urban region centered in Bangkok. The city is an important global city not only for the country but also for the wider region of Southeast Asia.

The country became a constitutional monarchy in 1931 but the government retained its Bangkok bias. Most political parties are based in the city; Bangkok elected officials may represent rural districts. The political, military, and intellectual elites live in Bangkok. The military, for example, plays a major role in Thai politics. There have been at least eighteen coups in the past seventy-five years, the most recent in 2006. The First Army, based in Bangkok, is seen as the vital route to power for ambitious soldiers. In order to attend elite schools and universities, it is necessary to live in the capital. Bangkok is the center of power for the military, monarchy, government, and bureaucracy (Askew 2002). In order to do business it is essential to be in Bangkok. Foreign companies need a Bangkok location in order to be effective. As the city grows migrants are attracted to the city. From 1960 the city's growth rate has been twice that of the rest of the country, fueled by rural to urban migration within Thailand and foreign immigration. Foreign migrants, initially from Burma, Cambodia, and Laos and latterly from Korea and China, are drawn to the city. The country has almost 2.3 million foreign migrants, the majority located in the capital city region.

Thailand's tremendous population and economic growth of the past fifty years was shaped by the initial urban primacy that limited alternative urban growth centers. Primacy skewed development and growth toward even more pronounced primacy. Rapid rural to urban migration, industrial takeoff and foreign investment in industry and services have all taken place in and through Bangkok. In a very real sense the country is one giant urban region. There are costs. A large number of studies have noted the environmental degradation that has marked urban growth in Bangkok (Rigg 1995; Fahn 2003). Concentrating economic and population growth in one relatively small area has created problems of air and water pollution that threaten the livability and sustainability of the urban environment. There is also the backlash from a national body politic divided between a prosperous city and an impoverished rural majority. When public unrest took place on the streets of Bangkok in late August and early September 2008, it was essentially a struggle between urban and rural political interests. The major political divide embodies the extreme primacy in the sharp division of political institutions and formal political representation into an urban middle class based in Bangkok and millions of poor rural people.

Middle-income Latins

There is a particular subset of the primate cases in Table 8.1. Ten are from Middle or South America—Argentina, Chile, Cuba, Costa Rica, Haiti, Jamaica, Nicaragua, Suriname, Uruguay, Peru—and nine of the ten are classified as middle-income.

These primates all share a colonial history. The centralizing element in the pre-Hispanic imperial and Spanish colonial systems (French, English/British, and Dutch in the case of Haiti, Jamaica, and Suriname respectively) set the conditions for subsequent growth. The largest capital cities were home to the elites, the population center of gravity as well as the economic hub of the national space economy. Primacy was reinforced as economies shifted from primary to secondary and then tertiary economic sectors, rural-to-urban migration increased, and foreign investment connected the local to the global. People, jobs, and investment moved to the major city. Llosa (2005) argues that corporatism, state mercantilism, and recent rounds of privatization have all reinforced the centuries-long persistence of social and spatial concentration of economic, political, and social power. Marked urban primacy is the spatial equivalent of the marked social supremacy of power, status, and wealth: the big cities have remained dominant just as the elites have maintained their privilege.

In some parts of Latin America, including Argentina, Mexico, Peru, and Uruguay, primacy is slightly down in recent years. In the case of Mexico, for example, Mexico City's population as a percentage of the national population increased throughout much of the twentieth century from a base of 7.4 in 1900, peaked at 14.0 in 1970 and steadily fell to 8.4 in 2005. However, in Mexico, as in many other Latin American countries, political power is still heavily concentrated in the capital, despite the growth in other urban centers, such as Monterrey and Guadalajara. This centralization of political power, which creates classic "rent-seeking," thus has a significant influence on shaping the spatial concentration of economic growth and development. As the economies of Latin America change, the heavy growth rates of the urban primates are leveling off. But even with declining urban primacy there are cases of continuing centralization at the wider regional level. In Peru, for example, the proportion of the national population in the greater Lima region increased from 27.6 percent in 1981 to 29.8 percent in 2005.

Urban primacy plays an important role in the creation of global cities. Consider the recent case of Santiago, Chile. In 2008 the metro population was seven million in a country with total population of 16.7 million. The initial urban primacy of a colonial system was reinforced by economic globalization. In the past decade over fifty multinational companies have located in the city, including Nestlé, Packard Bell, and Unilever. The city is a platform and export base for the wider South American market. Foreign investment from the US and Spain has committed the city to wider capital flows. The city also plays an important role in the cultural circuit of capital and the diffusion of managerial best practices. The city is a magnet for rural migrants in Chile but also for both unskilled and middle-income professionals from the neighboring countries of Bolivia, Ecuador, Peru, and to a lesser extent Argentina. The city is responsible for almost half of gross domestic product.

Urban primacy embodies and enhances the creation of global cities. Understanding the role of cities in national urban systems is crucial to explaining their position in the global urban hierarchy.

Low-income African countries

Sixteen African countries, all but one of them classified as low-income, have marked degrees of primacy. In order of primacy they are Togo, Nigeria, Uganda, Ethiopia, Guinea, Eritrea, Namibia, Mauritania, Côte d'Ivoire, Madagascar, Mali, Gambia, Sierra Leone, Tanzania, Gabon, and Chad. As with South American primacy, African primacy embodies colonial legacies and more recent rounds of economic globalization. Urban growth has increased dramatically in the large primate cities as well as in the non-primate cities. In the case of Côte d'Ivoire, for example, the primate city grew by eighteen times from 1960 to 2000, while the two non-primate cities grew by nine times. Even in the declining primacy of Ethiopia, the primate city grew by five times over this period while the two non-primate cities grew by seven times. The change in primacy rates is a function of the differential (high) growth rates of primate compared to non-primate cities.

New primates

Table 8.1 also lists a number of countries that we can designate as new primates. These include Armenia, Georgia, Latvia, and Serbia. They are national fragments from the former USSR and Yugoslavia. These regions' remaining large cities now stand out as primates. Previously they were merely part of a larger urban system. Such examples reinforce the importance of the scale of analysis and problematize the issue of urban primacy values in rapidly altered national urban systems. When the nation of the national urban system is radically altered, the resulting urban primacy figure may be more of a statistical anomaly. The new primates are in the stage of becoming rather than being, and their future evolution will depend on the interaction of global–national forces of urban agglomeration and dispersal.

Rich European primates: fragments of empire

Table 8.1 also contains a number of rich countries including Austria, the UK, and Greece while Hungary and Romania are at the higher levels of the upper-middle-income range (Figure 8.2). Urban primacy in all of these countries, apart from Greece, takes place against a background of slower population and urban growth than in Africa or Latin America. Part of European primacy pattern is an imperial legacy, the metropolitan end of the urban-empire connections that created colonial primate cities as sites of local control as well as metropolitan imperial hubs. Vienna, Budapest, and London were centers of far-flung empires rather than just capitals of individual countries. Nineteenth-century Vienna was one of the centers of political power in Europe, the capital of the vast Austro-Hungarian Empire. Vienna flourished as an imperial capital, the most important city in central Europe, but the loss of empire after World War I (1914–1918) reduced Austria to a small German-

Figure 8.2 Athens: one out every three people in Greece live in the capital city region.

Photo: John Rennie Short.

speaking state, and Vienna found itself a large city in a small country. Budapest was also the center of a much larger country and empire. In the Treaty of Trianon (1920), Hungary lost Vojvodina, Croatia, and Slavonia to the new state that became Yugoslavia and forfeited its coastline. In the north, Hungary lost to newly formed Czechoslovakia what is now Slovakia. In the east, Hungary had to cede Transylvania and the Banat to Romania. To the west, Hungary lost Burgenland to Austria. In all, Hungary forfeited two-thirds of its area. The end result was a large city in a much-reduced country. Nearly one-fifth of Hungary's population lives in Budapest, which is now more than nine times larger than the nation's second-largest city. In addition London was the center of a vast empire, formal and informal, that stretched around the globe. The legacy lives on in a huge city, filled with imperial archives, as capital of a small country (Figure 8.3).

Non-primates

At the other end of the continuum of primacy values are the non-primate distributions. Table 8.2 lists all those countries with a recorded primacy value of less than 0.9. Again it is difficult to discern a precise pattern. Low primacy is found in large countries as well as small, rich as well as poor, developed as well as developing. The case of Bolivia disproves the easy assumption of Latin American countries always having primate urban distributions. There

Figure 8.3 London: the city of London is at the hub of a 21 million population
        metropolitan region.

Photo: John Rennie Short.

are some very general trends. Size, either of area and/or population, is of some
importance in explaining low primacy as evinced in the case of Australia,
Canada, China, India, and the US. Yet no simple generalizations can be made
as low primacy is found in the Netherlands as well as India, the US as well
as Venezuela, Canada as well as Benin. If we look at just one example in
more detail a slightly different picture emerges.

Table 8.2 Low urban primacy and World Bank economic classification

Country	Primacy	World Bank classification
Benin	0.58	Low
South Africa	0.59	Upper middle
Venezuela	0.65	Upper income
Netherlands	0.70	High
Egypt	0.72	Lower middle
Australia	0.76	High
China	0.78	Lower middle
US	0.84	High
Bolivia	0.84	Lower middle
India	0.86	Low
Canada	0.89	High

Source: primacy calculated from data at www.citypopulation.de.

Figure 8.4 Brisbane: one of the big cities in Australia.

Photo: John Rennie Short.

Consider Australia, which according to its 2006 Census had a national population of 19.8 million. At first blush it appears a non-primate distribution because the largest city, Sydney, has a population of 4.11 million followed by Melbourne and Brisbane at 3.59 million and 1.76 million, respectively (Figure 8.4). The primacy value is 0.76. One could deduce that the urban system is evenly distributed. But on closer examination, as shown in Table 8.3, Australia is best depicted as a series of primate states in a federal system. Most states, except Tasmania, have very high degrees of primacy because the state capitals of Adelaide, Brisbane, Melbourne, Perth, and Sydney dominate the respective states of South Australia, Queensland, Victoria, Western Australia, and New South Wales. In South Australia, for example, in 2009

Table 8.3 Urban primacy in Australia, 2006

Country/state	Primacy
Australia	0.76
South Australia	33.48
Victoria	14.66
Western Australia	11.94
New South Wales	5.44
Queensland	5.32
Tasmania	1.13

Source: calculated from data from Australia Bureau of Statistics.

the capital city of Adelaide had a population of 1.19 million out of a total state population of 1.62 million. In Western Australia, Perth contained 74 percent of the state's total population of 2.25 million.

The Australian Federation of 1901 brought together a series of hyper-primate economies that the passing of time has done little to change. The major cities continue to dominate their respective states. The geography and (white) history of Australia play an important part. An antipodean gulag for the British soon developed into a colonial enterprise organized and structured through the port cities of the different states. A classic case of colonial spatial organization of a vast country was soon embedded into a series of primate states. The case of Australia shows that the sole reliance on national data abstracted from issues of scale, geography, and history has limited explanatory value. The overall primacy value for Australia is suggestive of non-primacy as if activities were evenly distributed throughout the urban system. Yet, when we look at the level of individual states, we see there is a marked concentration of activity, enterprise, and population in large cities.

Concluding comments: retheorizing urban primacy

The data analysis revealed the hyperprimacy of Thailand, Latin American middle-income primates, low-income African primates, as well as a scattering of primacy in richer countries. Hyperprimacy is more pronounced in smaller and medium-sized countries rather than very large countries, such as the US, Canada, or Russia for the simple reason that in larger, spatially dispersed economies there is higher probability of alternative and hence competing urban centers being established. Hyperprimacy is reinforced in centralized societies that can arise from traditional arrangements such as those in Thailand, as well as more recent colonial and postcolonial systems of control. The examples of Thailand and the UK highlight the problems with only using aggregate statistical analysis. By any standard economic development measure they are very different but they both share a history and persistence of marked primacy. The example of Australia also highlights the advantage of a more detailed approach. This is not to decry the search for broad patterns in the study of urban primacy. Rather, we need to refocus our theorizing on how city regions may or may not become the dominant spatial arrangement in a national economy. If the entire economy of a whole country is organized around one city region, then the result is primacy. This can occur in traditional societies such as Thailand, former imperial powers such as Austria and the UK, as well as in former colonies; in small as well as large countries; and in rich as well as poor countries.

Let us end this chapter with the rough outlines of a theory of urban primacy. The basic building block is the urban regional economy centering on one city. This city may become the center of a national space economy under a number of different circumstances including small to medium-sized homogeneous countries (Bangkok in Thailand, Addis Ababa in Ethiopia), imperial growth (London and Vienna in the case of the UK and Austria, respectively), and colonial domination and control (Montevideo in Uruguay, Lagos in Nigeria). Primate cities may also emerge in the case of the dismantling of empires into smaller national units as in the case of Yerevan in Armenia. Primacy may be reinforced or undermined during particular periods of economic change, including rapid industrialization, large-scale foreign investment, and pronounced rural–urban migration. Primacy is always in flux. While initial conditions of primacy tend to concentrate subsequent population and economic growth, declines in primacy may occur. In some cases it is simply the effect of scale analysis as the suburbanization of people beyond the narrowly defined city boundary makes it appear as if there is a reduction in primacy, whereas in fact it is the case of new metropolitan realities of a city region. In other cases genuine deconcentration may occur such as in Mexico, where growth rates in the primate city region are leveling off. In general, however, once primacy is established, while it may change its form, from city to city region primacy, it rarely disappears or even lessens substantially. Urban primacy is a classic case of the unfolding importance of initial conditions of urban economic geography influencing subsequent development through increasing returns from agglomeration.

Conditions of non-primacy can also be sketched. It is especially found in large, developed, and maturing economies where the sheer size of the country provokes the economies of agglomeration in more than one location. In the case of Australia, non-primacy turns out to be a national averaging of very primate states.

The key conclusion to draw is that the timing and form of national spatial organization and reorganization play an important role in the creation, promotion, and undermining of urban primacy. The unfolding relationship between a city region, state formation, and global incorporation is at the heart of the issue of urban primacy. And in turn, structures of national primacy feed into the project of globalizing cities, easing and reinforcing their insertion into the space of flows as hubs of economic and cultural globalization. In her discussion of globalization, Sasaki Sassen (2006) draws attention to changing global assemblages of territory, authority, and rights. I like the term global assemblages and suggest we hijack it to refer to the unfolding relations between city regions, nation-states and the global economy. An understanding of urban primacy can be recast as an example of these changing and constant assemblages.

Problems remain. Teasing out the causal connections between primacy and economic processes is difficult once we move away from the now generally agreed notion that urban agglomerations provide increasing rates of return. To what extent, when and where, can alternative agglomeration economies come into play in a national as well as the global urban system? And there is always the problematic conceptual connection between agglomeration economies and population increase. We can imagine a case where firms become more productive yet this does not necessarily lead to a population increase.

Productivity gains can be made through shedding labor. Drawing out the connections between economic growth and population growth, always factoring in the role of economic migration, is not an easy task. Does primacy drive the economy or does economic vibrancy drive population size? And then there is the recurring concern with the scale of analysis as city, city region and national scales overlap, coalesce and vary. Much work remains. We need to look further into the historical economic geography of nation-state formation and its connection with dominant city regions. We need to widen our attention to a deeper retheorization of urban primacy that is more sensitive to history, geography, and scale and concentrates on the shifting, siftings, and constancies in the city, state, and global assemblages of economic relations. Rather than rely only on aggregate modeling that reads off primacy from contemporary national characteristics, more theorized case studies are needed.

CASE STUDY 8.1 **London**

For centuries London was a central hub in the English then British Empire. Today Britain can be summarized as "London and the rest." The city and its extended metropolitan area squat over the nation just like a Third World primate city. The wealthy, the influential, the movers and shakers live in the city; it is home to Royalty, the political elites, those who manufacture the dominant forms of representation, and those who control much of the making and moving of money. This concentration of elites reinforces the centrality of the city in the national imaginary.

We need to be careful with terminology. While the City of London has a population of only around 9,200, and so-called Greater London comprises 7.5 million, the greater metropolitan region, which extends into most of the Southeast part of the country and stretches from the Wash to the Isle of Wight, has a population of approximately twenty-one million people, just over one in three of every person in the UK. People who live outside the city boundaries hold almost 20 percent of the city's 4.6 million jobs. London's primacy is undercounted because the data commonly used refer only to the population within the boundary of Greater London.

As early as 1906 H. G. Wells noted, in one of the more arresting urban similes, London, "like a bowl of viscid human fluid, boils sullenly over the rim of its encircling hills and slops messily and uglily into the home counties."

In her analysis of the city, Massey (2007) makes three points. First, in the past twenty years a financial elite was successful in representing London as a global city and this place making served class purposes. Second, she disputes the common assumptions that rising inequality is an unfortunate byproduct of a globalizing city and demonstrates that success and poverty are intimately related. Polarization is not accidental but integral to the making of the global city. Third, she explores London's role in the national space economy and criticizes the regional trickle-down hypothesis that London is the golden goose whose droppings fertilize the rest of the country.

London's dominance is not just the invention of the last two decades of neoliberal globalization. The longstanding and continuing dominance of London in the UK space economy has long fostered nationalist backlashes in the peripheries of the state. Scottish nationalism, for example, was always fueled by the overwhelming economic opportunities and massive government in London compared to the limited opportunities and frugal investments in Scotland. The distortion of the space economy was partly hidden in the Keynesian era by regional policies that sought to shift investment, especially public investment, away from the capital. Neoliberalism, in contrast, undermined the notion of regional equity and focused on global competitiveness. There was concentrated growth, especially in the financial services sector in London with massive deindustrialization in the periphery. In Britain the core–periphery is embodied in the distinction between London and the rest of the country. London like Paris is the national stage for representations of globalization and the national center for global representations.

London was not just the capital of the UK but a central node in a sprawling, vast, and worldwide formal and informal empire. The retreat from empire and decline of global economic dominance has left a big city in a relatively small country.

CASE STUDY 8.2 **Singapore city-state**

In the modern world there are only a few instances where the categories of city and state are essentially one and the same. This was a much more common occurrence in the early modern world when city-states were an important site of political identity and hub of economic networks. Assemblages of city-states such as the Hanseatic League were important forms of global connections. We get some idea of a global city network in the first city atlas, *Civitates Orbis Terrarum*, by Georg Braun and Frans Hogenberg, first published in 1572. It became so popular that by 1617 the work consisted of six volumes with over 363 urban views. In *Civitates* the city is both displayed and bounded. In almost all of the images, the city walls figure largely. Cities were often fiercely independent, the home to independent power centers, princes and prelates, guilds and town councils. Looking through the atlas at the many pictures of cities, one gains a very strong sense of cities as separate communities. Collectively, the images also indicate a world economy tied together in trade and linkages between urban centers. Aden, Peking, Cuzco, Goa, Mombassa, and Tangiers as well as other cities around the world are represented. While the cities are depicted separately, the effect of the compilation is to reveal a world economy of urban nodes and a trading world of connected cities (Short 2004c).

The modern world, in contrast, is one of nation-states. There are remnants of city-states and perhaps one the most influential is Singapore, a small island country dominated by a city. Almost five million people live in a country totaling no more than 710 square kilometers. The area became a British colony in 1824 and played an important role in the British Empire. It achieved independence in 1965. It is now an integral part of the global economy; it is the fourth largest financial center and is one of the busiest ports in the world. It is part of a complex network of flows. Jon Beaverstock (2002, 2007), for example, highlights its role in the career moves of executives in advanced producer services. It is one of the Asian tigers whose rapid economic growth was initially based on rapid industrialization. It is a culturally diverse city-state with four official languages recognized: Chinese, English, Malay, and Tamil. More than one in three of the population is foreign-born.

The nation of Singapore was in part imagined and constructed through and in changes in the urban landscape. Kong and Yeoh (2003) describe how the public housing allocation reflected ideologies of multiculturalism, street names were renamed in a postcolonial fashion, an Asian history was memorialized in heritage landscapes, and a performing arts complex was part of an attempt to construct a modernist city and project a global awareness.

Urban landscape transformation is a vital element in the global aspirations of this city-state. Han (2005) documents how the creation of institutional arrangement, regarding land development, the construction of the downtown, the building of industrial and business parks, and the provision of public housing were all tied to global city making. Chang et al. (2004) describe the developments along the riverfront as a "hyper-symbol" of global aspirations. Singapore is best understood as a global city-state (Old and Yeung 2004), that is constantly remaking itself to strategically locate into global flows. Ooi (2008), for example, considers the contemporary rebranding as a creative city in order to be part of the global creative economy. Singapore combines the aspirations of a global city, the trajectory of a developmental state within the institutional and ideological structure of the nation-state. City and state collapsed into one territorial unit imagined as an important hub in the global space of flows.

Part 3

City particularities

9 Alice Springs/Mparntwe

The postcolonial creative city

This chapter is another riff on the theme of globalization, the city, and modernity, but one that uses, at first blush, an unlikely case study. I draw upon a small town in the middle of Australia (Figure 9.1). It developed as a control center in the incorporation of interior Australia during the second wave of globalization in the late nineteenth century when the context was one of imperial authority, colonial control, and Anglo-Celtic Victorian cultural values. It re-emerges during the current wave of late modern globalization in new spatial assemblages of global nation-city and in a very different late modern/postmodern aesthetic. This case study also shows how the Late Modern Wave of Globalization shapes and is shaped in turn by changing aesthetics.

In the small Australian city of Alice Springs, also now known as Mparntwe, situated at the very center of the continent, art galleries selling the work of Aboriginal artists dominate the commercial heart of the city, along Todd Street and the adjoining streets. Despite the cultural and commercial centrality of their art, many Aborigines live in settlements located on the fringes of the town. In this chapter I will explore the complex factors behind both the dual naming of the place and the paradox of cultural prominence yet residential marginalization.

This theorized case study draws upon and extends our understanding of the postcolonial city. The term "postcolonial" originated in literary studies where it has three implications: a political position of anti-colonialism, an examination of the conditions of the postimperial world, and a category that critiques colonial domination (Kalliney 2002). There is now a vigorous school of postcolonial studies (Ashcroft 2006). Postcolonial urban theory concentrates on the conditions of postimperial places and is particularly concerned with the examination of colonial inscriptions on cities and the reinscriptions that follow decolonization. Yeoh (2001), for example, considers the issues of identity, encounters, and heritage. I extend this discussion by looking in detail at the spatial politics embedded, revealed, and contested in the transformation from the racialized urban space of the settler society to the more shared, yet still unequal, space of a postcolonial city in an ostensibly multicultural society.

Figure 9.1 Alice Springs/Mparntwe.
Photo: John Rennie Short.

There are two separate and distinct discourses in the urban scholarly literature, the postcolonial, that says little about urban economies, and the cultural/creative economy debate, which says little about postcolonial cities. This case study highlights and explains the connections between these two themes that are rarely conjoined. In this postcolonial, creative city the chasm between indigenous and colonist is not necessarily bridged as is usual, but stubbornly remains even as the mutual appropriations take place.

The reader can safely assume quotation marks around all words related to Aborigine in the rest of this chapter as the term is so heavily freighted with numerous complex, contested, and inconsistent usages. The descriptor is confused and confusing, inconsistently mobile. Terms associated with "Aboriginal" change routinely, sometimes claimed and reclaimed, other times rejected and devalued, all part of a revolving politics of identity. Aboriginality is not a slowly evolving biological fact: it is cultural construct forged from a subaltern position, shaped by a dominant non-Aborigine culture, and always in relation to changing political discourses.

Pre-colonial and colonial

Alice Springs/Mparntwe is located very close to the geographic center of Australia. The town sits on the northern side of the McDonnell Ranges, a dramatic feature of parallel red quartzite ridges that reach up to 3,000 meters at their highest point and stretch for 400 kilometers across the stark, eroded desert landscape. The range is punctured by erosion gorges where fresh water is seasonally available. The town sits next to one, Heavitree Gap, four more are close by, and their English names are Pine Gap, Simpson Gap, Emily Gap, and Jessie Gap (see Table 9.1).

Table 9.1 Dual place names

Arrernte	English
Akeyulerre	Billy Goat Hill
Anthwerrke	Emily Gap
Atnelkentyarliweke	Anzac Hill
Lhere Mparntwe	River Todd
Mparntwe	Alice Springs
Ntaripe	Heavitree Gap
Ntyarlkarle Tyaneme	Unnamed ridge

The town has at least two histories, a very long pre-colonial and a much shorter colonial. The first history comprises the stories of the Arrernte people who moved into the region around 35,000 years ago. Keen (2004) identifies seven main types of cultural-ecological regions in pre-contact Australia. Mparntwe/Alice Springs is firmly located in the central dry desert category. Because of the poor soils and low, erratic rainfall, population densities were low, approximately one person per 80–200 sq. km. The total Arrernte pre-contact population is estimated between 8,000 and 10,000. Food supply was scarce and tool development was limited by the need for portability. The landscape was envisioned and managed as a complex system of land titles based on lineage and family and conception. Responsibilities were connected to wider kinship links through both paternal and maternal lines. Responsibility for "managing" specific sites often lay in the hands of senior elders, both men and women. Most sites are gender-specific in their ownership and responsibility patterns. The complex interlocking titles were less monopoly controls and more relational and totemic, allowing individuals and groups a broad range of claims, responsibilities, and bargaining options that enabled long-term occupancy of a harsh environment with irregular water and limited food sources. A cosmology connected people to the land: it was dominated by an emphasis on ancestral traces in the country, the flattening of time between the time of the ancestors (*altyerre*) and the present and the importance of specific land features, such as waterholes, watercourses, trees, hills, and gorges as both vital resources and totemic sites. Space rather than time was the predominant axis of cosmological meaning. The cosmology bound people to the land in intricate webs of meaning that also sustained long-term economic usage. For a range of recent work on the Arrernte, see Austin-Broos (2009) and David (2008). For a more reflective piece on the production of knowledge about the Arrernte, Sam Gill (1998) provides a fascinating account of successive descriptions of tribal beliefs and practices as constructed by anthropologists and other non-Arrernte observers. In these foundational, classic texts the Arrernte are always viewed and described from a non-aboriginal perspective; they are represented not as they are to themselves but as they appear to non-indigenous onlookers.

According to traditional stories that have been recorded, the landscape around Mparntwe/Alice Springs was shaped by ancestral figures including caterpillars, dogs, and wallaroos as well as two sisters and uninitiated young boys (Brooks 2003). The creation story begins at Anthwerrke/Emily Gap where the caterpillar beings ate their way through the Ranges. Three species of caterpillar were involved, but the most important is the Yeperenye who shaped the topography as they moved cross the landscape. As they crossed the dry riverbed of Lhere Mparntwe/Todd River, a caterpillar form was created as a small ridge, Ntyarlkaerle Tyaneme. Other ancestral beings also played a role. A wild dog, for example, shaped Ntaripe/Heavitree Gap in the course of fighting an adversary. After the struggle he metamorphosed into a boulder at Akeyulerre/Billy Goat Hill. According to Arrernte cosmology, the whole area around Mparntwe/Alice Springs is rich in stories and song lines. The dry bed of the Lhere Mparntwe/Todd River is a particularly significant site. In this Arrernte history the past is conflated into the singular, present landscape that requires respect and celebration in ritual and song. This precolonial history left the continuing legacy of a connection with the land, which, as we will see, was reconfigured as a more commodified form in the colonial/postcolonial context of the encouragement of indigenous art production.

There is another view of the land derived from the more recent colonial settling of the region and the country. In this rendering the landscape is more inert, more a container for human activity, a backdrop for economic and cultural events. There are areas of special significance within it: Anzac Hill/ Atnelkentyarliweke, for example, is a non-indigenous sacred space as memorial to Australia's war dead, but elsewhere it is very largely a landscape evacuated of any deep spiritual meaning and profound cultural resonance. This history does, however, have the equivalents of important ancestral beings. In this case it begins with John McDouall Stuart (1815–1866). Born in Scotland, he moved to Australia and worked as a surveyor, but soon became involved in explorations into the interior. He was the first officially recognized white man to travel across the continent north to south. In 1860 he led his third expedition into the continental center. He named the McDonnell Ranges after the Governor of South Australia. His own name lives on: Alice Springs was originally called Stuart; it changed to Alice Springs only in 1933, yet there is still a Stuart Caravan Park, a Stuart Lodge Guesthouse, and a Stuart Highway that, following his track, runs all the way from Adelaide to Darwin though Alice Springs/Mparntwe.

In 1870 the South Australian government agreed to build an overland telegraph line to Darwin that would link up with one of the earliest global communications networks, the global telegraph system. The route across Australia followed Stuart's tracks most of the way. The telegraph needed repeater stations along its route to boost the signal. It took seven hours to transmit a message from Australia to London. In 1872 a repeater telegraph station was established just outside present day Alice Springs/Mparntwe.

The River Todd was named after the Superintendent of the whole project, Charles Todd, while the springs were named after his wife Alice.

The two different histories—pre-colonial and colonial—now collide, as Alice Springs becomes a center of control and domination. The telegraph station was the first of subsequent surges in urban development as gold and rubies were found, pastoral farming was developed, and the town became an important military base. The Arrernte had no title to the land in the English-based legal system. In 1874 an area of 25 square miles around the telegraph stations was simply annexed. Across central Australia, pastoralists, white farmers raising cattle, began a process of land dispossession. Aborigines resisted the process. In 1874 Aborigines attacked a cattle station, killing two white men: in retaliation at least fifty Aborigines were killed. The white public authorities sought to pacify the Aborigines, but there was an enduring conflict between the local people and the settlers' cattle since the livestock competed with traditional game for scarce sources of food. The Aborigines killed the cattle for food. Constables and white stockmen hunted down "cattle killers." The town was at the center of a bloody struggle as dispossessions and subsequent guerilla reprisals were met in turn with brutal government action and killings (Kimber 1991).

The Coniston Massacre was the culmination of an escalating conflict over land and food, sparked off by a specific incident over sexual favors. Starvation was pronounced in Aboriginal communities as continuing land dispossession took away their traditional sources of food and sustenance. On August 7, 1928 pastoralist Fred Brooks was killed after a deal with two men for the sexual favors of an Aboriginal woman went awry. Patrols sent out to capture the offenders killed seventeen Aborigines in the first expedition, fourteen in the second, with a final total of 100 innocent Aborigines killed. As Bruce Elder (2003) has well documented, what happened in Coniston was not an isolated event in white–Aboriginal relations along the frontier. As the episode still resides within the arc of living memory, the legacy endures; there are many Aboriginal people in central Australia who remember or have family members who remember the Coniston incident.

In 1927 the town was made the administrative capital of central Australia and soon after a railway line linked the town to Adelaide. As Alice Springs/ Mparntwe grew, it functioned as a colonial city of control and was imagined as a white town, viewed in a colonial gaze that dismissed the Aboriginal presence. In 1928 the town was declared a prohibited area for Aborigines: they required a pass before entering the town. These restrictions in effect initiated the local unfolding of the white national imaginary of Australia. A regime of effective government regulation and surveillance tightly controlled the Aborigines (Markus 1990). From 1911 to 1953 Aborigines were moved and relocated to reserves and missions with no heed given to their concerns or wishes. From 1953 they were classified as wards of the state and required official permission to marry, leave reserves, dispose of property, drink alcohol, or open a bank account. These restrictions were only lifted in 1964 when Aborigines could own land and legally drink alcohol. Even until the late 1960s Aboriginal children were often removed from their families and placed in orphanages or white foster homes.

Control was also effectively maintained through the allocation of food and other items. Tim Rowse (1998) documents how in central Australia, throughout the nineteenth and first half of the twentieth century, the authorities issued rations of food and blankets to indigenous people. Declining game and limited access to traditional lands put the indigenous people on the threshold of famine and starvation. Government-supplied rations, distributed by white ranchers, missionaries, and police, were an important element in the effective control over the indigenous people.

The postcolonial interleaving and combining of the indigenous and non-indigenous is evident in more recent historical narratives. Liam Campbell's (2006) biography of Darby Jampininpa Ross, who survived the deaths of his family at the Coniston massacre to become a drover and prospector as well as painter and land rights advocate, shows how one man negotiated the opportunities and constraints set by the dominant culture while also expressing and articulating his own culture. The book *The Town Grew up Dancing* celebrates the life and art of Wenten Rubuntja, an Arrernte man who grew up in Alice Springs/Mparntwe (Rubuntja 2002). The story follows his development as an artist, land rights activist, and community leader, providing a portrait of the changing nature of white–Aborigine relations in the town. The book is also interesting for its collaborative nature; the text was produced from interviews, the authors are listed as "Wenten Rubuntja with Jenny Green with contributions from Tim Rowse" and the text contains large sections in Arrernte.

The shift of Aborigines from ration recipients to citizens also involves the transformation of Alice Springs to Alice Springs/Mparntwe and from white town to shared space. Four developments have transpired in this awkward, problematic, continually evolving, postcolonial space in the past thirty years. The first was the formal recognition of an Aboriginal presence in the city, the result of a downward cascading political effect from changes in federal legislation as well as of an upwelling of local resistance and contestation. This change was embodied in the legal claim to land rights in the town. The second was the development and formal recognition of "town camps." The third was the growing issue of the politics of and regulation of urban public space, and the fourth was the development of a cultural economy based on Aboriginal art. Let us consider each in turn.

Land rights and the meaning and naming of space

Since the 1970s a new national imaginary of Australia as a postcolonial multicultural society has slowly developed (Hamilton 1990; Gelder and Jacobs 1998; Povinelli 2002; Anderson and Taylor 2005; Goot and Rowse 2007). Enduring inequalities and a continuing dominant-subaltern relationship mark this representational space. And although Aborigines are constructed and located in this space by the dominant white society, they now play larger and more complex roles as numerous national, state, and local pieces of legislation have been enacted. The most important policies that affected Aborigines in Alice Springs/Mparntwe were connected to land rights. The 1976 Aboriginal Land Rights Act recognized land claims of indigenous communities only on Crown (i.e., public) land that was not leased. The underlying assumption was that the land prior to British annexation was *terra nullius*. In 1982 three Torres Strait islanders, Eddie Mabo, David Passi, and James Rice, successfully challenged this concept. Two of the justices, Deane and Gaudron, noted in their ruling that Australia's treatment of Aborigines was "a national legacy of unutterable shame"(*Mabo and Others v. Queensland* 1992). The follow-up legislation, the 1993 Native Title Act, created a working legal framework, and it was under this piece of legislation that application was made in 1994 for native title to lands and waters in and around the town. The applicants were representatives of the Arrernte people and in particular three branches of this tribe, the Mparntwe, Antulye, and Irlpme. An application was lodged in 1994, and the case, *Hayes v. Northern Territory* (1999), went to trial beginning July 1, 1997 and concluding in February 1999. In May 2000 the Federal Court decided that Arrernte native titleholders should retain their rights for most of the reserve, park, and vacant crown land in the town. The legal process was not simply an adjudication of property rights; it was a complex melding of the traditional and the modern. On the one hand it allowed indigenous land claims into an English-inspired legal system. On the other it connected traditional titleholders to commodified land markets and state bureaucracies. The *Hayes v. Northern Territory* (1992) decision accepted that Arrernte people lived in and around Alice Springs and identified landholders, who held the group rights by virtue of their age and lineage. The ruling gave the land to traditional title holders, along with a set of new rights: the right to be acknowledged as traditional owners, the right to use the natural resources, the right to make decisions about the land, the right to protect places, the right to manage the spiritual forces and safeguard their cultural knowledge associated with the land. The legal ruling was a complex acceptance of modern land property rights as well as recognition of pre-modern group rights and associated spiritual and cultural dimensions to land holdings.

The judgment created new subjectivities and new forms of Aboriginal–white relations. The subjectivities included the legal acceptance of Aboriginal landownership titles. The term Kwertengerle can be translated, roughly, as "spokesperson" or "manager" of a piece of territory. The judgment accepted the legitimacy of this position and gave it legal significance and bargaining power. A traditional Arrernte role was now embodied in the formal legal system. In the process both the belief system, because it is now articulated in non-traditional setting with elements of incipient commodification, and the legal system, since it now has to deal with issues of cosmology, are subtly changed and transformed. It is not simply that traditional owners are now legal owners; a more complex transformation is wrought as traditional owners articulate their requests through property concerns while property rights are given a cosmological interpretation.

New bureaucratic identities were also created in the wake of the court ruling. The Lhere Artepe Aboriginal Corporation (LAAC), composed of representatives of the three Arrernte groups, was formed in 2002. The LAAC now plays a significant role at formal events. Official visits from state, federal, and international representatives are formally welcomed by representatives of LAAC, and greeted with the performance of subtle, and not so subtle, enactments of the multiple identity of the important place and the role of indigenous landowners. The LAAC also provides protocols for visitors to the town as well as general cultural and educational information about the town. The town is encoded in an Aboriginal presence. For example, in 2006 members of LACC were involved in the painting of municipal waste bins in "traditional" Aboriginal designs. In these and other ways the multiple meanings of the town and the ownership of these meanings are performed, invoked, and embodied. The official sign that welcomes visitors to the city now contains the multiple meanings in referencing the traditional owners of the land alongside Alice Springs Town Council.

New Aboriginal–white relations were re-established in general and especially with regard to the conflict between land use changes and sacred sites (Figure 9.2). While there had been some recognition of Aboriginal sensitivities before the land rights legislation—a plan to build a dam across the Todd River to create a lake was abandoned in 1984 in the wake of complaints that it would seriously disturb a sacred site, and a flood mitigation dam proposal was quashed in 1992 for similar reasons—not all conflicts were avoided as in the case of Ntyarlkaerle Tyaneme, the sacred shape created by the caterpillar beings. The Northern Territory Government constructed a road to link a new casino to the town. In December 1983, despite the long discussions with traditional custodians of the land, the government drove the road through the "tail of the caterpillar." The white name for the road is Barrett Drive; many of the local Arrernte people refer to it as Broken Promise Drive. Today, in and around the city, an official recognition of sacred sites means that they are listed, documented, and protected much more than they once were. The modern town is filled with the recorded evidence of the ancestral beings.

The ruling also created a changed urban identity. The city's occupancy of Arrernte territory is now signaled at official ceremonies, as well as by urban signage. The town is no longer just Alice Springs and no more just the expression of white explorations; it is Alice Springs/Mparntwe, and it encompasses reminders of Arrernte creation stories. But note I use the specific term Arrernte not the more general "Aboriginal." There are tensions between Arrernte title-holders and non-Arrernte visitors from other Aboriginal communities that the lands claims case exacerbated. More of this later.

Figure 9.2 Sacred space within the town limits.
Photo: John Rennie Short.

Residential segregation

Despite the early, explicitly racist legislation denying access to the town, there was always a persistent Aboriginal presence. The members of the Arrernte people maintained their links with the land while other groups were attracted to the town by the demand for labor and the rations issued by the police. In the 1930s there were estimated to be around 400 Aborigines in Alice Springs area living in informal settlements clustering around the town and along the bed of the Todd River. Traditional ceremonies were still enacted and performed, even in these marginalized circumstances (Rubuntja 2002). The Aboriginal population continued to grow as people moved into the town from the surrounding reserves and missions, often places of authoritarian compulsion. The population also increased after Aboriginal stockmen working in the pastoral sector, using their new non-ward status, successfully campaigned in 1966 to have their wages increased. Pastoralists responded by sacking many of them, and some ended up in Alice Springs/Mparntwe. Many Aborigines moved to the town in search of jobs and services. Pushed off the land, they often had nowhere else to go.

By 2006 the population of Alice Springs/Mparntwe was 23,891 with 20 percent or 4,494 people considered to be indigenous. This contrasts with the nation as a whole, where the indigenous population constitutes less than 3 percent of the total population. The indigenous people in the area include people traditionally associated with Alice Springs/Mparntwe as well as groups whose traditional lands are located throughout central Australia. The more detailed 2001 Census reveals a marked difference between indigenous and non-indigenous populations in Alice Springs/Mparntwe. The former tend to be younger and poorer. While the modal weekly income category of indigenous households was less than $399, for non-indigenous it was between $1,500 and $1,999. The disparities in living standards in Alice Springs/Mparntwe are replicated across the nation: While there has been some absolute improvement from 1971, advances have been slow, and there has been little improvement in rates of unemployment and health (Altman et al. 2004). Unemployment is three times higher for indigenous peoples compared to non-indigenous while median household income is only half. Life expectancy for indigenous men and women is 59 and 65; the respective figures for the non-indigenous are 77 and 82 (Pink and Allbon 2008).

Aborigines live in the town proper as well as in the town camps, areas of informal settlements that grew on the margins of the town (Figure 9.3). The 2001 Census showed 716 in camps and 2,150 in the town of Alice Springs. In his survey of town camps based on the census data, Will Sanders (2004) reports the town camp population is more similar in profile to that of remote communities in central Australia than the indigenous population of Alice Springs proper, in that they are younger, poorer, less educated, and consist of larger families; 76 percent are not in the labor force compared to 40 percent and 22 percent of the indigenous and non-indigenous populations of

Figure 9.3 Dwelling in a town camp.
Photo: John Rennie Short.

Alice Springs respectively. And compared to the indigenous people in the rest of the town, most of the town camp population speaks an indigenous language at home.

A more detailed survey of the town camp population suggests the Census undercounts the population. Foster et al. (2005) estimate the permanent town camp population closer to 2,000 living at high densities with an average of ten to sixteen people per house. Almost a third of the town camp population is in transit from remote communities, elevating the population estimates to over 3,000 at any one time. The town camps are linked to remote communities throughout the region; people travel from the outback into town and back again to maintain family links, do business (in both senses of economy and culture), to visit health and other centralized services. People move to and from the town camps and remote communities for ceremonial purposes and duties such as "sorry business" (the funereal and grieving rituals that play such a big part in central Australian Aboriginal culture), to sell art, to collect royalties, and to visit health centers. In effect, there are two Aboriginal populations in Alice Springs, those who live in the town—much closer in profile to the non-indigenous population—and the town camp population—closer in character-istics to people in the remote communities.

Town camps are relatively small communities made up of members of the same language group. There are now twenty camps located around Alice Springs/Mparntwe and the principal language groups are in order of size: Arrernte, Walpiri, Anmatyerre, Pitjantjatjara, Luritja, and Kaytetye.

The camps are located in the direction of the traditional country with Kaytetye in the north, Pitjantjatjara to the south, Walpiri in the northwest, Luritja in the west, with the eastern and western Arrernte appropriately located.

We can picture the evolution of town camps as a three-stage model. Stage one is of colonial occupation as traditional Aboriginal camping grounds became incorporated into the urban fabric of a white power structure and function as rations depots and labor camps. Stage two is of enforced segregation from 1911 to 1964 when Aboriginal residence and movement are very tightly controlled. Camps are discouraged by the racist ideology but the ideology is trumped by economic necessity. Aborigines provide much needed labor, and the camps, now defined as "fringe" dwellings, function as a nearby labor pool. The fringe dwellings became town camps as more people move in and dwellings became more permanent structures. In this stage there is a continual tension and struggle between, on the one hand, the official policy of dispersion and, on the other, labor market requirements of white employers and the resistance of town camp dwellers. Stage three begins in 1964 with legislation that overturns the policy of the town as prohibited areas to Aborigines. The more affluent and better-educated Aborigines move, tentatively at first, into a formerly white town. A bifurcation of the Aboriginal community develops based on ability to work in and through the non-Aboriginal society. Those with more power and ability to navigate the "white" system, tend to move into the formal town neighborhoods. Those less skilled and more connected to the outlying indigenous communities reside in the town camps.

In 1974 town camp residents formed Tangentyere (an Arrernte word meaning "working together") to demand better services. This council, officially recognized in 1977, now provides a range of services including transport, welfare, policing, and housing assistance. Along with the Alice Springs Town Council and LAAC, Tangentyere Council now provides another level of governance and urban identity.

This close inspection of residential patterns reveals a more complex picture of Aborigines in the town. The notion of a singular Aboriginal community hides the fracturing into camp and non-camp Aboriginal communities, a residential divide that embodies a growing cleavage between those directly linked to traditional lands and practices and another, more adept at navigating through and in the dominant culture. One group is still imprisoned by Aboriginality in a dominantly non-Aborigine society, while the other can and does use Aboriginality, in this dominant white but increasingly multicultural imaginary, as a platform for advancement. As Aborigines are incorporated into the national imaginaries and political discourses, those more able to negotiate this process can become relatively more successful, indeed often using their Aboriginality as a bargaining chip, a source of legitimacy, to achieve greater socio-economic advancement. For others, more closely tied to non-English speaking communities in remote Australia and marginal Australia, Aboriginality continues to be a source of inequality. The distinction is not always so stark. There are town camp dwellers that do bridge the

Aborigine/non-Aborigine divide. But in general the town camp dwellers are those whose Aboriginality is less of an advantage in a changing imaginary and more of a continuing disadvantage in a neocolonial society.

The politics of public space

The black–white relationship is problematic in Australia, as elsewhere, but in Alice Springs/ Mparntwe the relationship is more intense. From its origins as a racialized urban space, the town has watched a complex racial drama play out. Since 1964 all formal restrictions on Aborigines in the town have been lifted; however, given the huge disparities in income, living standards and often codes of behaviors, the sharing of public space was always a tense affair. Even into the late 1970s Robyn Davidson (1980, p. 8) could describe the city as "a frontier town, characterized by an aggressive masculine ethic and severe racial tensions." Issues of behavior in public space center on public alcohol consumption, and Aborigines are often demonized as lazy drunks. In 1979 the shop owners on Todd Mall petitioned the Council to curb loitering in the streets, a response to the perception that the presence of Aborigines frightened away customers. Public drunkenness was indeed an issue as people would come into the town from remote communities, especially those where drinking was prohibited, sleep in the sand riverbed of the dry Todd River, and drink in public. In 1981 a strychnine-laced bottle of sherry was left in a public space; it was picked up by some Aborigines who drank it in the Todd River riverbed. Two died and six were made seriously ill.

As the town becomes more explicitly Aboriginal, the politics of urban space become more complex. Accommodation jostles with incarceration, greater accessibility with ongoing constraints. Much of the tensions over the use of urban public space continue to center around public drinking. Alice Springs/ Mparntwe is now a dry town where it is illegal to consume alcohol in public spaces, although it is not uncommon to see empty beer bottles piled up against a "no alcohol" sign. Issues of public behavior are not simply a white concern. The LAAC issues protocols to Aboriginal visitors reminding them not to camp on sacred sites, not to "humbug," that is, beg for money and cigarettes, and to avoid alcohol and fighting. Tangentyere Council also publishes social behavior rules for "bush" visitors staying in town camps, which remind them to respect Arrernte sacred sites, especially prohibiting camping and drinking in these sites and other public spaces.

In Alice Springs/Mparntwe there is tension in the exploitive reliance on an Aboriginal creativity. Since issues of authenticity dominate the purchasing of art, Aboriginality is an important element in the commercial viability of the art. Yet controlling Aboriginal behaviors in public places is considered essential to the commercial viability of the town. Many white tourists are frightened off by too many Aborigines and most prefer the exotic at a commodified distance rather than as fellow users of urban public space.

The demand for "proper" behavior, however, crosses racial lines. The debate is not just between Aborigines and non-Aborigines: there is also a tension between the indigenous visitors from outlying areas who come in to drink and socialize, and the indigenous, long-term locals. Numerous Aborigine organizations in the town promote public space decorum. In this and as in so many realms, there is no singular Aboriginal identity or shared set of interests.

Indigenous art and the cultural economy of a postcolonial city

In the commercial heart of the town that runs along the central retail spine of Todd Street and its adjoining streets there are galleries and stores selling "Aboriginal" art; they dominate the prime commercial locations. This concentration results from a complex interaction of government intervention with indigenous cultural expression, the changing supply–demand interactions of the indigenous art market, and the agglomeration economies identified in the standard economic geography literature, but with a neocolonial/postcolonial twist. From its role as a colonial command center, the town has become more of a postcolonial collaboration center for indigenous art production and sale; however, the colonial legacy persists through the unequal exchanges.

Federal and state governments initially stimulated the Aboriginal cultural economy as a way to develop the tourist industry and generate employment opportunities for indigenous communities. The Aboriginal Arts Board was established in 1973 to encourage and foster indigenous art and crafts. Arts centers were established in many remote indigenous communities and art advisors were employed to help in selling and marketing art produced in the local community. Initially, arts centers were used in the communities to give people something to do, to foster community awareness and social bonds and only partly as income-generating centers. By the mid 1990s, in the wake of uncertain and fluctuating federal and state aid, arts centers were perceived more in terms of income-generating sources. Despite very high social needs, remote indigenous communities have few sources of making money. Selling art was vigorously promoted to generate local revenue streams. From its small beginnings the Aboriginal art market has grown rapidly (Altman 2005). Arts centers have expanded, from sixteen in 1980 to just over 100 by 2008. There are now between 6,000 and 7,000 indigenous people making art and artifacts for sale. Combined sales estimated at $A 2.5 million in 1980 were worth $A 300 million by 2006 (Kremmer 2006). Art sales now form a significant source of revenue for Aboriginal communities throughout Australia and especially in central Australia. Some of the art is sold direct from the arts centers but much of it is funneled through dealers and galleries in Alice Springs/Mparntwe.

While government initiatives have prompted the indigenous art market, indigenous people themselves have vigorously embraced the opportunities. The best-documented case of a bottom-up initiative is Papunya Tula (Perkins and Fink 2000; Bardon and Bardon 2004). In 1971 a young Sydney teacher, Geoffrey Bardon, went to work in the government settlement of Papunya, approximately 250 km northwest of Alice Springs/Mparntwe, established in 1959 as a central location to assimilate the tribes of the central desert. He developed good rapport with the local people, some influence with senior men, and was able to encourage and promote the production of art. A mural painted on the white exterior school walls by a variety of people was the first expression of the artistic capacities of the local people. Later, paintings were made, first on composition board and plywood and then on canvas, using natural pigment as well as synthetic paints, employing designs tradition-ally drawn on the body or in the sand. The paintings were visual repre-sentations that were spiritual maps, since people painted the sites for which they had spiritual responsibility. The paintings told stories of the creation of the land. The central desert art movement was born. A regular painting group coalesced around thirty men, including Clifford Possum Tjapaltjarri and Uta Uta Tjangala. The men soon formed the Papunya Tula Artists Pty Ltd as a communal artist cooperative to market and sell its paintings. Papunya Tula annual art sales now average close to $A5 million. The success of Papunya was a model for other communities.

Art of the central desert area of Australia cannot be easily subsumed under one reading. Howard Morphy (2008) writes of Aboriginal art as intercultural. It is an art that fulfills many functions: as a form of communication to insiders about their connection to land, as a form of communication with outsiders, as an expression of group identity and political rights (paintings and other forms of art that represent territory have been used to promote land claims), and as an important source of monetary income (Watson 2003; Genocchio 2008). Aboriginal artists draw upon their embellished material culture and cosmology and transform it into commodified art. Eric Michaels (1994), in one of the more sophisticated analyses, draws attention to what he terms the "fantasy of authenticity" and argues for the artists as willing agents of modernity, revealing but also hiding and deconstructing the sacred and secret elements in their work. Too much of the debate focuses on the authenticity of the art; so that the issue becomes measuring the distance between the "real historical past" and its more commodified market-orientated present. The result is often one of disappointment and, paradoxically, a constant privileging of the influence of non-Aboriginal power on Aboriginal art. Michaels suggests a change in focus away from authenticity toward authority, a shift that promotes a greater appreciation of the agency of Aboriginal artists in both their ability to navigate an intercultural space and in their role of creating and hiding as well as replicating and disclosing. In a similar vein, Myers (2002) follows the paintings as they move from central Australia through dealers, museums, and critics to become high art, but also shows how they relate to the painters' own hopes for recognition. This more complex pers-pective places Aboriginal art in the mainstream of modernist aesthetics and postmodern sensibilities rather than in the marginal, indigenous, and pre-modern. It strips away the condescending connotation of "Aboriginal" often implied in the use of the term "Aboriginal art" and it allows us to see "indigenous" art as a creative and collaborative experience, and Aboriginal artists as innovative producers of cultural commodities.

The term "Aboriginal" or "indigenous" art assumes an unchanging aesthetic, which is far removed from the reality of today's fast-changing world that requires equally rapid changes in its representations. The term also assumes a one-way relationship as indigenous artists make art. In reality the art produced is mediated through art advisors, gallery owners, and collectors who inform, guide, and help shape the final products. Contemporary Aboriginal art is now finely tuned to the changing consumer demands of both national and overseas markets. The art is produced as a commodity to be bought and sold. And in this commercial nexus we come to the importance of the postcolonial city.

Alice Springs/Mparntwe is the main urban center for the indigenous population scattered throughout central Australia in remote communities. It is an important cultural center. Dunbar-Hall and Gibson (2005, p. 189), for example, write of its importance in the contemporary indigenous music scene and note how, "Alice Springs is emblematic of a political economy of cultural production, where symbols of the exotic and ancient create tourist markets, distinct from the cultural forms and expressions that are popular in everyday life among local Aboriginal people." As an important hub in the indigenous art market, the town is situated at the heart of a complex intercultural space. It is home to formal institutions such as the headquarter site of Desart, the Association of Central Australian Aboriginal Art and Craft Centers. There are now forty-two arts centers throughout central Australia affiliated with Desart. Full time indigenous artists often live in the town for short or long periods, some even permanently. Alice Springs/Mparntwe acts as a magnet for artists living in the outlying and distant regions of central Australia who come to visit friends and family, visit health facilities or do business with the range of organizations, including the Central Land Council and Native Title Representative Board. Artists often combine these visits with the opportunity also to market their artwork.

The town is a center of aesthetic innovation and cultural creativity. The prevailing discussion of such worldwide urban phenomena tends to concentrate on the advanced producer services in cities. Richard Florida (2002), for example, attracts considerable attention as a popularizer of the idea of a *creative class*, defined as the collective talent behind cultural economies. His work also has attracted significant criticism. Peck (2005) provides a sustained criticism embedded in a more general condemnation of neoliberal urban programs. In this particular urban location, the collective talent is indigenous people with a unique relationship to the land that produce art representing and depicting this relationship, in association with art advisors, sellers, and brokers of art. The production of Aboriginal art is as much a creative collaborative process as designing iPhones or making movies.

This local creative outburst is a product of changes at three distinct but interrelated scales. At the global scale are the aesthetic sensibilities of the post-colonial as a condition of postmodernity. The very condition and expression of postcoloniality involves a revaluation of the indigenous, a celebration of the native, and a commodification of the tribal. Postcoloniality involves a fundamental cultural and economic reassessment of the indigenous and the native. At the national level, there has been a shift in Australia from the dominant aesthetics of an Anglo-Celtic settler society to a more multicultural sensibility that celebrates and rewards indigenous art. The international and national are connected. A consistent theme of recent international understandings and representations of Australia is an emphasis on the indigenous character and legacy. Gordon Waitt (1999) demonstrates how official tourist promotions exploit the image of the indigenous to signify ecotourism and to suggest an entry into an experience of a pre-modern wilderness. The encouragement of the indigenous is a particularly important element not only in terms of internal politics but also in the selling of Australia's national image abroad. The indigenous element gives Australia an important and distinctive niche in the global tourist and art trades.

At the urban level, the city acts as liminal space between the indigenous and non-indigenous, the transformational site where material culture is turned into commodified art and a transmission point in the supply chain that transforms, packages, and explains the material culture of the desert into the art for the metropolitan center. The city's postcolonial transformation hinges around the creative industry of "Aboriginal" art. Art production is now one of the single biggest economic sectors in Alice Springs and its surrounding regions. The town is home to over fifty art sale outlets. They range from the souvenir shops that sell small "tourist art works," often for no more than $A50 to more up-market art galleries selling works in the range of $A50,000 (Figure 9.4). The old souvenir shops once dominated. Many of these small-scale, family-owned enterprises sold Aboriginal souvenirs including wooden crafts, spears, didgeridoos (not an instrument indigenous to this part of Australia), and boomerangs. As the art market took off, they added canvases to their inventory. The Australian Aboriginal Dreamtime Gallery formerly situated on Todd Mall originally sold wooden crafts but started selling canvasses in the 1990s. Some of the souvenir shops developed into gallery spaces specializing in the fine art markets.

Mbantua Gallery opened in 1992 and employed just one person. By 2008 it had expanded, taking over the space formerly occupied by a dry cleaners and a Thai restaurant. It now employs twenty people and incorporates the site formerly held by the Aboriginal Dreamtime Gallery on Gregory Terrace. The gallery owner has invested substantial sums building a cultural museum in the space. Mbantua sells the work of 250 artists specializing in the work of people from Utopia, a region northeast of Alice Springs. There are also galleries such as Gallery Gondwana, situated on Todd Mall, which has always catered more to the fine art market, encouraging the work of younger artists

Figure 9.4 Art gallery.
Photo: John Rennie Short.

and specializing in specific artists, such as Dorothy Napangardi and Mitjili Napanangka Gibson. Gondwana, opened in 1990, regularly hosts annual exhibitions such as *Divas of the Desert* and solo shows. Annual turnover is over $A2 million. In 2004 a companion gallery was opened in Sydney. There are also arts center galleries. The most famous is the Papunya Tula Gallery selling work by members of the Papunya Tula cooperative. From 1973 it used to inhabit a small dusty dismal building in Todd Street before moving to a more spacious location right in the middle of Todd Mall in 2004. It now has the "white box" feel of an upmarket gallery.

The location of the galleries is important. The prime sites are on Todd Mall, an open-air pedestrian thoroughfare, and are now inhabited by Gondwana Gallery, Mbantua, Papunya Tula Gallery, and one of the oldest commercial galleries, Aboriginal Desert Art Gallery. The only Aboriginal-owned gallery is situated half a mile away from this central hub. To establish a gallery that sells Aboriginal art requires a complex and culturally specific mix of capital, access to credit and familiarity with business practices and the regulatory system that few central Australian Aborigines, at this time, possess.

Galleries open and close, wax and wane, and relocate. Closures of recent years include the Aranda Gallery, Original Dreamtime Gallery, and Red Sand Gallery. In 2008 alone, two new galleries opened, Kuruwarrin on Todd Street Mall specializing in central desert art, Palya Proper on Gregory Terrace. The former Yanda Gallery became the abpg Gallery while Boomerang Gallery, which sells work from Utopia moved to a location between Palya Proper and abpg. The Arunta Art Gallery, one of the oldest, played a significant role in the early development of Aboriginal art. Its owner is a redoubtable woman who figures in Bruce Chatwin's bestselling 1987 book, *The Songlines*. The gallery's space is welcoming, with wonderful books for sale and still some art material but there is little new inventory.

The galleries in Alice Springs both retail art to buyers but also wholesale to other galleries in Australia and around the world. Galleries not only sell art but also are sites of production. Many of the galleries have painting rooms and even painting houses offsite where indigenous artists work while they are in town. There are a number of circuits that link producer to consumer. There is the formal process whereby arts centers sell the combined work of the artists to galleries and dealers. There are freelance buyers who visit the art-producing centers to buy art direct from both the artists and arts centers. Mbantua Gallery staff drive to the Utopia area every week to drop off art supplies and purchase art.

The concentration and clustering of art galleries provides an example of a Marshallian node that maximizes labor pooling, input sharing, and innovation spillovers—but with a colonial legacy. On the one hand, as predicted by the standard economic literature, the dense network of proximate galleries allows gallery owners to share information on artists, prices, customers, new techniques, and new opportunities. For the artists the range of galleries allows a concentrated opportunity to shop their work around. Galleries compete for the work of blue-chip artists whose work commands many thousands of dollars. Buyers of indigenous art are given an opportunity to compare prices and artwork all within a half-mile square block. Innovation spillovers, for example, are exemplified in the rapid changes in style that are quickly incorporated by artists from different regions represented in different galleries. These sudden shifts are indicative of a more functionally connected system than previously believed. The linkages are also sustained through personal connections. It is not uncommon for arts center coordinators to work in art galleries, some even setting up their own galleries. People who work in one gallery can easily move to another, while artists can and do present their work in different galleries. The art gallery scene is a tight nexus of interconnecting commercial interests and personal ties.

On the other hand, there is still a colonial legacy that shapes economic transactions. The interactions between producers and sellers are often marked by inequalities. The most obvious are between, on the one hand, indigenous artists, some unable to speak English and often in desperate need of money, and, on the other, sharp gallery owners and canny dealers. Individual gallery owners and dealers try to skim off the most prized work by the big name artists while arts centers want to promote the work of all their artists. As with all art markets, renowned artists can bargain for higher prices. The more successful artists will sell their work directly to galleries and dealers. There are also independent entrepreneurs who specialize in taking orders from buyers and then arranging with the more prominent artists to come to town and do the paintings. This contract system has been the subject of much scrutiny for its abuses. After a journalistic exposé of downright forgery and of how badly some artists were treated (Rothwell 2006), an Australian Senate (2007) inquiry led to numerous proposals, including tougher powers to combat forgers and unethical behaviors. The Australian Competition and Consumer Commission (2007a; 2007b) now prints leaflets for artists and arts centers as well as for consumers of indigenous art. The colonial legacy lives on in the poor treatment of many indigenous artists in the commodified art market.

Alice Springs/Mparntwe is now an important site of the production and sale of indigenous art. Commodified indigenous art is an intercultural product that relies on the special knowledge, in this case the spiritual responsibility for sacred sites, of the artist to make the art, and of experts such as arts center coordinators, art gallery workers, and art dealers to make sense of the art for metropolitan audiences. Both sets of knowledge-based professionals require some measure of trust in order to interact to produce and sell indigenous art. As in all economic transactions, levels of trust are heightened by regular face-to-face contact. The town concentrates these knowledge-based experts, heightens the possibility of intercultural narration, and creates opportunities for face-to-face contact. Information flows both ways as art experts guide artists to produce more sellable products in a constantly changing environment, and artists provide some of the background to the special, partly revealed knowledge that informs the work and gives it the necessary allure of indigenous uniqueness so attractive to many buyers. With its clustering of galleries and concentration of knowledge experts, the town is an important epistemic community that creates and packages indigenous art.

Indigenous art, its production, sale, and display are now a significant element in not only the formal and informal economy of the town but also how the city represents and sells itself. Art is celebrated in the city. The city is home to festivals such as the annual ten-day Alice Desert Festival. An integral part of this September event is the Desert Mob Art Exhibition. Established in 1991, the annual three-day affair now includes a daylong symposium and two days of art sales that bring together artists and their work from over forty arts centers throughout central Australia and buyers from around the country and the world. The desert art aesthetic is encoded throughout the town. Visitors entering the airport terminal building are met with a giant painting, *Yuparli Dreaming*, by Eunice Napangardi. The Araluen Arts Center's exterior is covered with the work of indigenous artists, and even the rubbish bins in the downtown are emblazoned with the iconography of indigenous designs. There are also art-producing centers within the town, including the Tjanpi Desert Weavers; Irrkerlantye Arts, for Eastern and Central Arrernte people living in Alice Springs and surrounding areas; the Ngurratjuta Iltja Ntjarra (Many Hands) Arts Center, which promotes the work of the figurative water colorists, landscape painters; and Tangentyere Artists, which sells and promotes the work of around 400 artists from the town camps.

Alice Springs/Mparntwe is a creative city at the intersection of Aboriginal art production and national and global consumption. It sits in that complex, awkward space of the shift from colonial to postcolonial through the commodification of the precolonial.

Conclusions

The shift from colonial control center to postcolonial city involves four inter-related processes: land rights negotiations, residential segregation, the use and regulation of public space, and the development of a cultural-creative economy based on "Aboriginal" art. Exploring the paradox of cultural prominence and economic importance of Aboriginal art with the continuing residential marginalization of many Aboriginal inhabitants highlights continuances and ruptures of the postcolonial city. The city resonates with the tensions between the postcolonialism of new national imaginaries and the continuing colonialism expressed and embodied in the commodification of "Aboriginal" art. A consideration of this final process provides an intriguing twist on the usual suspects of postcoloniality. The transition from colonial to postcolonial shows signs of rupture as well as continuity.

In this particular city, a new property rights regime and the development of cultural industries based on the work of indigenous artists informs the transformational shift from colonial control center to a postcolonial shared space. The continuing colonial legacy is expressed in the issues of behavior in urban public space, continued residential segregation and marginalization of indigenous people, and in the unequal transactions that mark many of the dealings between indigenous artists and non-indigenous cultural entrepreneurs. The city is a complex meld of the continuing colonial and the emerging postcolonial.

The city also reflects how changes in national imaginaries filter down to local places, in this particular case, how a dominant colonial Anglo-Celtic imaginary gives way to a fuller expression of an Aboriginal authorial presence. In turn, this new Aboriginal authority prompts a revaluation of indigenous art, itself now both a construction for a cosmopolitan audience and an assertion for an indigenous audience. The process also goes in the other direction from the local to the national and even the global. The emergence of the desert art movement, with Alice Springs/Mparntwe at its center also influences the character and form of a new national imaginary and shapes a new global representation of Australia as a country with an indigenous presence. It is now possible to buy "Aboriginal" motifs in a range of objects from scarves to screen savers, from carpets to dishcloths, and from t-shirts to sweaters. Qantas cabin crews have ties, skirts, and dresses with "Aboriginal" designs drawn from the motifs of indigenous artists from the central desert region. This is part of a wider appropriation of Aboriginal art motifs in the material and cultural representation of Australia. Major banks cover their ATM machines in desert dot painting motifs; the Australia Post issues stamps with the image of paintings from Papunya Tula. Corporate headquarters in Melbourne and Sydney, Parliament buildings in all the states, as well as federal government offices in Canberra display prominently the work of Aboriginal artists, and Australian embassies around the world show the work of indigenous artists in permanent installations and occasional exhibitions. There is a colonization of the Australian imagination as Aboriginal aesthetics inform contemporary design, public space, and national representations in the wider world. The indigenous colonization of the Australian aesthetic imagination hints at the complex connections between the national and the local, Australia and Alice Springs/Mparntwe. This case study is suggestive of a broader role of a city in the representation of a country and the more general notion of urban–nation–global linkages and flows moving in either direction.

For most Australians living in the large cities on the water's edge of the continent, Aborigines are imagined citizens, rarely seen or encountered in person. In Alice Springs/Mparntwe, in contrast, the Aboriginal presence is large and constant. In this city the new national imaginary is unfolding in all its tensions and difficulties. It is a place where reconciliation is being worked out in a lived place rather than the easy space of rhetoric. The indigenous and the non-indigenous meet not only in the commercial space of innovation and creativity but also in the experiential lived space of the city. As a newly emerging postcolonial city it is still in the process of becoming. The town is not simply a place where Aborigines now live and their artistic work is presented but where multiple and competing identities, of Aborigine and non-Aborigine, Arrernte and non-Arrernte, native title holders and non-title holders, long-term residents and short-term visitors (also both Aboriginal and non-Aboriginal), are given shape and life and traction. In the relational urban space of the emerging postcolonial, new forms of local and national identity are shaped, formed, and contested.

In this particular city setting, the constituent elements of the term of postcolonial become unglued. In terms of an urban cultural economy based on indigenous art, the emphasis seems more on the "post" of the postcolonial. Artists and communities do receive both money and recognition. However, in terms of the unequal exchanges that often mark the indigenous art industry and the continuing residential segregation, especially between the town camps and the rest of the city, the picture is still more "colonial" than post. Postcolonial cities embody the flow of change along with the fixity of persistence. At the heart of this particular postcolonial city is the new aesthetic and commercial centrality of Aboriginal art with the durability of residential segregation/marginalization of many Aborigines. There is no easy path from colonial to postcolonial. It is not an easy passage but a rough process, full of enduring inequalities, filled with paradoxes, embodying old conflicts and generating new tensions as well as new hopes.

I first wrote these lines on a Sunday morning in September 2008 while nursing a coffee and looking out on the Todd Mall. Artists were selling their art in the street, gallery workers were opening the galleries and visitors to the town were out sightseeing and shopping. Some people were getting their morning newspapers; others were going to church. I could see black and white, old and young, indigenous and non-indigenous, men and women from town camps selling art, tourists from around the world taking photographs, people going about their individual business, yet collectively sharing the same space, a place full of hope and disappointment, rage and compassion, anger and love.

CASE STUDY 9.1 **National imaginaries and the city**

The representation of cities is bound up in national imaginaries that in turn are shaped in and through the lived experiences of cities. Nations, according to Benedict Anderson (1991), are communities as much imagined as lived. National imaginaries are rarely coherent, consistent, or stable. The dominant national imaginary in Australia has shifted from celebration as an Anglo-Celtic colonial outpost to the more uncertain embrace of a postcolonial, multicultural society. A major element in this shift is the changing role assigned to indigenous people. In the nineteenth century their presence was considered a problem to be overcome, and throughout much of the twentieth century the solution was a policy of willful neglect, active marginalization, and assimilation. As the Chief Protector of Aborigines in the Northern Territory said in the 1930s, part of this policy aimed to "breed out the color" (Johnston 2008).

"Aborigine" is too fraught with imprecision to be truly definitive. Yet in the last thirty-five years the term increasingly designates an important component of a distinctly Australian national identity. Such imaginaries take on more authoritative shape as they are enacted: witness the presentation of Australian national identity at successive Olympic Games. At the 1956 Melbourne Games, indigenous motifs were incorporated into designs, but an Olympic postcard showing a naked female Aborigine throwing a spear was classically "Aboriginalizing," employing the dominant representation from a white settler society perspective. Aborigines were of marginal interest to an Anglo-Celtic nation under a southern sky. At the 2000 Sydney Games, Aborigines were not only given prominence in Australia's enactment of national identity to a global audience, but they also were allowed to present themselves. The gold medal success of Cathy Freeman, as both an Aborigine and an Australian, seemed to signify a more mature postcolonial society.

National imaginaries' shifting paradigms are recorded in historical writings. Historians such as Henry Reynolds now tell history from the other side of the frontier (Reynolds 1981, 2001). A major television documentary of Aboriginal history, first released in October 2008, is significantly entitled *The First Australians*. This change is also embodied in public pronouncements and policies. It was only in 1967 that Aborigines became citizens rather then merely wards of the state.

The 1976 milestone legislation, of the Aboriginal Land Rights Act, began a process of recognition of indigenous native land claims. When a 1997 government report detailed the heart-rending stories of children taken away from their families, the so-called stolen generation, a National Sorry Day was declared, and over 300 events were held throughout the country. A million people signed Sorry Books. On February 3, 2008 the recently elected Prime Minister Kevin Rudd offered an official apology.

The construction of a postcolonial identity is not shaped by equal partners but clawed at by unequal participants. Like mismatched wrestlers, these participants do not maintain unyielding differences but rather flexible relational identities that shift strategically in response to changing opportunities—sometimes defensively, at other times aggressively.

National identity is shaped by how Aboriginal practices are encoded into law. Gelder and Jacobs (1998) offer the example of how Aboriginal beliefs affect national identities, showing how Aboriginal claims for sacred sites were specifically codified and then diffused into wider discourses of identity and meaning. Indigenous cosmologies were mapped onto contemporary social debates, changing each in turn to create a more complex identity that unsettles the settler society, creates an anxious postcolonial racism and involves the cultural appropriation of Aboriginal sacredness by non Aborigines. The "white" response is strung out along a continuum from a fetish for the indigenous to the politics of resentment against Aborigines. However, Goot and Rowse (2007) distinguish the difficulty in confidently establishing the public's response to indigenous issues by effectively demonstrating how opinion polls rarely reveal the full complexity of white public opinion.

Even when indigenous issues, such as land rights, become part of the national discourse, they are filtered through the dominant/subaltern relationship. Povinelli (2002) explores how white recognition highlights certain forms of Aboriginality while excluding and marginalizing others. Multiculturalism is not an acceptance of Aboriginality but a creation and foregrounding of certain forms of aboriginality. The shift from a colonial to postcolonial national imaginary in which Aboriginality plays an important role is still an unequal process, reflecting the hierarchic relations of power, that involves the creation of Aboriginal subjectivity in a society still marked by the white dominance of a settler state. A new Australian imaginary is not a mere widening of sources to incorporate an uncomplicated acceptance of the indigenous but another fold in a complex colonial relationship between colonized and colonizer. This new fold creates new spaces of opportunities for accommodation and recognition as well as cooption and marginalization.

Much of the scholarly focus on national imaginaries has skewed towards cultural representation and discursive space. Yet it is important to remember that transformations in national imaginaries connect with urban economies and local politics. There is an inevitably uneven geography to the adoption, diffusion, and enactment of changing national imaginaries, and tensions between the rapidity of national shifts and the relative inertia and/or relative "stickiness" of more local practices.

National imaginaries not only filter down to local places, they emerge from these places. The emergence of the desert art movement, with Alice Springs/ Mparntwe at its center, is an important part of the new national imaginary, and global representation, of Australia.

CASE STUDY 9.2 **A question of method**

This chapter grew out of an obsession with Australian art. I first visited Australia in 1985. I lived in the country for two years and became fascinated by its art, a subject I have written about and returned to in a number of different publications (Short 1991). I returned in 1997 and began paying annual visits to Alice Springs/ Mparntwe. My role was initially less as a researcher and more as someone simply wanting to lean more about art of the central desert. In the beginning I had no agenda to "do research" or "write up my results." I spoke with gallery owners and artists. The town's inhabitants are used to "blow-ins," people and especially academic experts who visit once, maybe twice at most, and locals, Aborigine and non-Aborigine, develop a certain reserve and contempt for the passing transient visitor. I was in a unique position as a very regular visitor, keen to talk and buy, a privileged participant, but a participant all the same, whose interaction was not shaped initially by the distorting nature of an a priori research agenda or the chilling need to codify these encounters into a research topic. From America but with a Scottish accent, an art buyer but also an academic, I was difficult to place and categorize, which made more open discussion all the more possible. As a buyer I was given access to the process perhaps not available to only researchers. I was not a dispassionate observer but an active participant. After nine consecutive years of visits and many art purchases I conducted an informal survey of all art galleries in the town in 2006. This consisted of semi-structured interviews of owners and workers, a final sample of twenty-two, asking about their personal narratives, as well as more general questions about relations with artists, arts centers, and the dynamics of art purchases and art sales. I ensured confidentiality in order to obtain information. In such a small world individuals are easy to identify even with pseudonyms. Over the years I also spoke with fifteen artists who either lived or worked regularly in the town. Again confidentiality was assured. Two difficulties arose. The first was a general reluctance of gallery personnel in the immediate wake of the numerous high profile exposés of the industry. My repeated visits and regular purchases allayed most but probably not all of the fears. The second was my difficulty speaking with artists who did not speak English. I have very limited skills in the languages of the central desert and so was almost entirely reliant on translators.

The result is a set of observations from a person who visited the town on a regular basis over many years and talked with lots of people connected to the art market; it arises from someone who listened and mapped, wrote and observed, bought and conversed. As with all fieldwork I am indebted to the comfort of strangers. I interacted on a commercial and personal basis before interviewing gallery owners, artists, and local bureaucrats. I also used the marvelous Alice Springs Collection at the local library as a principal source for documents. As well as reading and writing, I also talked with as many people as I could over the course of what is now twelve annual visits. I interacted with senior women in the Lhere Artepe Aboriginal Corporation who provided me with lots of information, insights, and opinion. In a natural gesture of reciprocity that fieldworkers often use, I used my rental car to drive them to outlying properties as well as around town. There are, of course, many remaining issues of difference, otherness, and inequality that this kind of work continues to raise and provoke. However, I am hopeful that there is, as Popke (2009, p. 81) reminds us, a new understanding of ethics that is more about "enhancing and celebrating our immersion in Being."

I make no claim that this form of fieldwork authorizes knowledge. The world is not an exhibition, being there is just being there, not a privileged site for knowledge production.

---

10 Shanghai

The reglobalizing city

Shanghai is at the leading edge of the Third Urban Revolution, on the cusp of the Late Modern Wave of Globalization and one of the main centers of an emerging metropolitan modernity. The recent experience of Shanghai highlights three things: the process of urban reglobalization as a city reconnected to the global economy and networks of cultural and political globalization after a period of disconnect; the rapid urbanization of a developing economy; and the metropolitan modernity of a city involved in repositioning itself as a self-consciously modern and global city (Figure 10.1).

While sharing the same dizzying rates of growth and spectacular urban expansion as other Chinese cities, Shanghai is also a Chinese city with a more cosmopolitan past. The city has no long history as an important imperial administrative center. Unlike Beijing whose historical roots lie deep in Chinese history, Shanghai is a relative newcomer to the status of major city. For centuries it existed, at a bend in the Huangpu River, as a trading and market center. It grew after 1684 when restrictions were lifted on sea trade. By the 1830s it was a prosperous trading port, but with few foreign traders. All this was to change with the 1842 Treaty of Nanjing. The growth of the city, and its ambiguous role in the Chinese imaginary, was the result of a weakening of the Qing Dynasty in the face of aggressive European colonial power. The British, in particular, were keen to expand their commercial empire, and China, then as now, was considered a vast and lucrative market. Tea was shipped to Britain and silver was sold to China. But the most profitable trade was opium, grown in India and consumed in China. In the early nineteenth century, the trade increased fivefold. As addiction increased, the trade boomed, as did the flow of wealth from China to British interests to pay for the drug. British companies benefited as an addictive and destructive drug was foisted onto the Chinese. The Chinese authorities tried to resist. In 1838 more than 20,000 chests of opium were destroyed in Canton (Guangzhou) and the port was closed to foreign shipping. The British retaliated. In the First Opium War from 1840 to 1842, Chinese resistance was easily overcome by British military power. In the Treaty of Nanjing, signed in August 1842, the island of Hong Kong was ceded to the British, compensation was paid for the destroyed opium totaling almost half a billion dollars in today's US currency and five treaty

Figure 10.1 Shanghai: the city grew and continues to grow beside the Huangpu
River.

Photo: John Rennie Short.

ports were opened in Canton, Xiamen, Fuzhou, Ningbo, and Shanghai. It was
a national humiliation that marked the end of Chinese national sovereignty
over all its territory, resulting in the direct imprint of foreign power and
inaugurating the spectacular growth of Shanghai.

Treaty ports, where foreigners were allowed access to markets, residence
rights, and freedom from local authority, signified a loss of national sovereignty and a forcible entry into global trading patterns. Sixteen more Chinese
treaty posts were established by 1860 with a total of fifty by the end of the
century. Foreigners could reside in these treaty ports, where they were exempt
from internal tariffs and not subject to the national laws and regulations.
In 1854 US warships under the command of Commodore Perry "opened
up" the Japanese port of Shimoda; the same year a Russian fleet "opened up"
Nagasaki. In 1876, the Treaty of Ganghwa "opened up" three treaty ports
in Korea for the Japanese. Forcing treaty ports became the preferred form of
neocolonial dominance. Treaty ports were imposed by outside powers. They
signified the end of economic and political sovereignty. They also transformed
local places. Nowhere was this transformation so large and complete as in
Shanghai.

The colonial city

As a treaty port Shanghai now had an ambiguous position as it was located
within China but exempt from the rule of the Chinese authorities, the Qing
dynasty until 1912 and the Chinese Republic until 1949: it was a Chinese city
controlled by foreigners. Shanghai occupied a liminal position between the
fringes of national state power and the sharp edges of colonial economic and

cultural penetration. The city grew because of the lucrative opportunities provided by this ambiguous space. It became an important port and trade center importing goods and ideas from around the globe and exporting goods and materials to world markets.

The British were first granted 140 acres along the foreshore of the river where they established their customs houses, clubs, and banks on a narrow embankment, the Bund, which was eventually to feature the tallest buildings in all of Asia. Then came the Americans, the French, and later the Japanese. The International Settlement consisting of the British and American Concessions stretched along the Huangpu River; the French Concession was to their south. The old walled Chinese city was now surrounded by the foreign settlements. In 1854 legislation allowed Chinese nationals to buy property in the foreign settlements. Migrants from the Chinese countryside flocked to the city in search of employment opportunities and to escape the ravages of civil wars racking the country. The rapid population growth created a huge demand for land and inevitable price increases occurred. Fortunes were made in land and property speculation, and not for the last time.

The city's growth was a function of the incorporation of China into the world economy. Banks and trading companies were established to lubricate the foreign trade. The Hong Kong and Shanghai Bank was headquartered in the city—it still exists as a corporate entity, now called HSBC. Shanghai became one of the largest shipping ports in the world. It exported food and raw material from the Chinese interior to world markets. Tea, silk, and raw material were shipped through the city. Trade increased because of the space–time convergence initiated by the railway and the increased use of steam navigation, which reduced the time it took to transport goods. The city also imported items from the wider world, including consumer goods, manufactured goods, and opium. Shanghai was at the entry point of China's global trade. The city became responsible for 40 percent of all external trade in the entire country. It was a purely commercial venture, economics and profits trumping everything else. As one foreign trader noted, "our business is to make money, as much and as fast as we can" (quoted in Dong 2000, p. 15).

The city became China's major industrial center with mills, factories, chemical plants, and shipyards. Chinese, British, and Japanese capital was invested into factories along the Huangpu River. Textiles were the leading sector employing thousands of people. By 1927 there were twenty-four cotton mills in Shanghai. In the 1930s there were 75,000 workers in textile factories in the International Settlement and 73,400 workers in the Chinese municipality. In both places combined, there were 1,257 factories. Individual cotton mills employed between 1,000 and 4,000 workers; foreign-owned factories tended to be larger than Chinese-owned factories (Hinder 1942). The city was now an important node in world trade, its economy intimately connected to global trading patterns and flows. It grew when exchange rates were favorable as with a rise in global demand for raw materials and/or a fall in the value of gold which cheapened the cost of imports. The city moved to the rhythm of global trade and the beat of international finance.

The city's spectacular economic growth was reflected in population increase and spatial restructuring. In 1860 the city's population was close 300,000. By 1900 the city's population approached one million and by 1930 was close to three million, including 100,000 foreigners. Almost one million people lived in the foreign settlement areas, and its main through route, Nanjing Road, was a commercial artery of the entire city. The street was the scene of a famous incident that embodies the complex nature of the city. When Sikh policeman reportedly killed a Chinese laborer in a Japanese factory, Chinese students marched in protest. On May 30, 1925 the ringleaders were arrested and held in a police station on the Nanjing Road. Crowds demanded their release but Sikh and Chinese police, led by an Englishman, fired on them, killing four outright with five dying later. Strikes and riots were called in protest in Shanghai and across China.

At its precommunist peak in the 1930s Shanghai was the fifth largest city in the world, a cosmopolitan city with at least sixty different nationalities including Russians escaping the Revolution, Sikhs from India, Iraqi Jews, Vietnamese, Koreans, Americans, and Europeans.

Places and practices of conspicuous consumption soon developed in this economy of quick profits and easy money. Bars and clubs, racecourses, palatial residences, and impressive commercial buildings all sprang up. Shanghai continued to expand as its economic growth pulled in rural migrants from around China and attracted foreigners from all over the world.

Although there was little direct social interaction between the Chinese and non-Chinese and what did occur was always within a colonial and neocolonial power structure, there were transfer encounters. For example, many of the factories employed Chinese engineers who were trained abroad and returned with their newfound industrial skills. The city was a place where transnational encounters, albeit filtered through a colonial relationship, created a modern capitalist economy and a modern city. The city was not only a beachhead of economic globalization; it was also a site of cultural globalization as ideas and practices impacted Shanghai as they spread throughout the global urban network. A distinctive architectural form, *shikumen*, emerged in the narrow alleys of high-density housing, opening out to more open courtyards, a hybrid of European and Chinese designs. In the foreign areas neoclassical buildings competed with the new modern buildings that referenced New York as much as China. The Chinese descriptive for skyscrapers, "magical big buildings that reach the sky," evokes the sense of wonder at this urban modernity. Shanghai visual artists such as Xu Beihing (1895–1953) and Guan Zilan (1903–1986) transformed traditional Chinese painting by working in Western styles. A hybrid form of "Shanghai modern" was evident in graphic arts and new designs for rugs and furniture.

The rapid industrialization and urbanization created, as it did elsewhere in the modern world, the context for working-class movements. As in Great

Britain earlier, and later throughout Europe and North America, the rapid increase in factory labor allowed workers to see, feel, and experience their class position and collective identity. The city and the factory, as Marx had first noted, could turn peasant isolation into working class consciousness. From 1919 there was marked politicization of workers, especially in the vast textile factories. A class in itself became a class for itself as a new class identity was forged in the urban crucible of a rapidly growing and rapidly industrializing city (Honig 1992). An official survey of the industrial scene in the International Settlement alone listed 170,704 people employed in 3,421 factories (Hinder 1942). The Chinese Communist Party was established in Shanghai in 1921 in a *shikumen* housing block in the French Concession; the location of its birth a metaphor for the hybrid nature of metropolitan modernity. As an emergence of national class identity, the party was both communist and Chinese and had to compete with and sometimes utilize the more parochial interest of the numerous and important native-place associations that continued to play major roles for rural migrants in the metropolis. In the city migrants from specific regions and even particular villages formed mutual help associations (Goodman 1995). It was only in the industrial city context that these could be transformed into wider links with fellow workers from other parts of China. The sparks of international communism ignited in the textile factories of Shanghai.

The Chinese transliteration for modern, *modeng*, was first coined in Shanghai. The city was a site of modernity, both cultural and economic. It was a place of European imprint but also of Chinese accommodation, incorporation, and resistance to modernity. Lee (1999) writes of the flowering of a new urban culture in the city, a "Shanghai Modern," in which the modern was both consumed and produced.

Shanghai also inhabited a complex discursive space, represented, on the one hand, as a cesspool of wanton debauchery by both Western missionaries and Chinese nationalists, and, on the other hand, as a neon-lighted, department-stored, clean-watered, electrified, telephone-connected, modern city with lots of economic opportunities. The city was an important node in a global economy and important hub in the transmission of ideas and practices of metropolitan modernization. China's passage to modernity was like a powerful jet of water forced through the narrow pipe of coastal cities and especially the big coastal city of Shanghai.

Early interpretations of the city as entirely the result of foreign imprinting, what was done to the Chinese by the foreigners, are now replaced with a more complex reading of the Chinese embracing of modernity without legitimating colonial subservience (Yue 2006; Knight; Yee 2007; Chan 2010). Gaulton (1981, p. 40) describes how "Shanghai's middle and upper classes cultivated a distinctly Europeanized cultural style." In literature, architecture, film, theater, and political ideologies, the Chinese of Shanghai were as much makers as subjects of an urban modernity.

The communist city

Shanghai became a communist city in 1949. On May 25 of that year troops of the People's Liberation Army (PLA) marched into and took over the city. Their triumphant arrival marked the end of a difficult period in the life of the city after years of civil unrest, war, bombings, terror, and a brutal Japanese occupation. In 1927 Chiang Kai-shek at the head of a 100,000 man army took over the city. He launched what is now know as the White Terror against the communists and organized labor. The Chinese civil war between Nationalists and communist forces lasted from 1927 to 1949. In 1931 the Japanese Army invaded Manchuria and began their territorial incursions into China. The Japanese took over the city on December 7, 1941, the same day as they bombed Pearl Harbor. With Japan's unconditional surrender in 1945, the city passed into Nationalist control. But not for long. The victory of the PLA meant that the city would no longer be a colonial city or a Nationalist city, but a communist city.

There was a prediction in the early years of communist rule that "the communists will ruin Shanghai and Shanghai will ruin the communists" (Gaulton 1981, p. 36). Shanghai's commercial resilience was quickly in evidence. On the very same day that the PLA entered the city, counterfeit communist money was already being printed in the city. Photographs by Cartier-Bresson illustrate an article published in *Life* magazine in October 1949. One photograph shows peasant soldiers looking awe-struck into a department store window front displaying refrigerators. Their slack-jawed wonder was soon replaced by a quick transformation of the city. The *Life* article noted that within months of the communist takeover,

> more than 800 restaurants have closed down . . . tea houses are now also gradually vanishing . . . Scores of cinemas are also undergoing difficult times . . . Bars, coffee rooms and other public places are closing down one after another. Shanghai is changing and under the new Shanghai there will be a new social order.
>
> (Boyle 1949, p. 141)

It was not simply a replacement of one elite by another—although in the French Concession senior communist officials did take over the plush residences of the rich foreigners—rather it was the construction of a different city, one reimagined as a site of proletarian production.

There was a complex relationship between Shanghai and Chinese communism. The party emerged from the city, but was largely a rural-based entity during the long struggle with the Nationalists. The Party drew upon a vast reservoir of peasant discontent. The communists were taking control of a cosmopolitan capitalist modern city. Some believed that communism would destroy the old city while the city would undermine the best intentions of communism. In the early years it was the former that triumphed. The communists imposed their power on the city as on the rest of the country. Land and business were nationalized, and the labor movement was strictly controlled. After years of conspicuous consumption came the years of austerity. People were urged to eat less than an ounce of rice day. Conservation was less a lifestyle choice than an act of political survival. Emphasis was on the utilitarian rather than the luxurious. The simple cotton uniform replaced Western fashion styles. There was a mass mobilization of residents to provide public services such as street cleaning. The cosmopolitan pleasure-loving city became a more austere production center. A Chinese communist city replaced a cosmopolitan capitalist city.

The earliest economic plans of the Party aimed to build up heavy industry. In the very first Five-Year Plan, heavy industry was promoted in the city. Steel production, an iconic sector in the life of early communist regimes, signaling modernity and industry, was encouraged. The city was reimagined as a proletarian production center to replace the colonial cosmopolitan metropolis. The light industrial base of the city was shifted toward heavy industry such as machinery, metallurgy, and shipbuilding. Even in this new economic order the city's comparative advantage was evident. The old industrial base and pools of skilled labor ensured that the city was one of the more efficient economic performers in the Chinese national economy. And even the skills of the artistic community were employed as the vibrant print culture of the colonial city was transformed into the aggressive and lively revolutionary posters of the communist city. By 1978 it was responsible for one-third of the country's entire exports. Even as late as the early 1990s, in every passenger train coming into the city an announcement was made to passengers that they were entering "the largest industrial city in China."

In the very earliest years of communist control the city grew quickly. State policy toward Shanghai was to encourage its heavy industry. Migrants from the rural areas flocked to the city as factories provided employment opportunities. The city population quickly grew from four million in 1950 to almost 6.5 million in 1960, a staggering 50 percent increase.

Then the city was viewed with suspicion. The Party was worried by the disparities in wealth and economic capacity between the coast and the interior. The government favored building up the industrial production of small and medium-sized cities in the interior. Rapid big city growth worried the communist leadership; it siphoned off peasants from the countryside and could concentrate resistance to Party rule. The rural peasantry was idolized by party ideologies as the source of a communist purity. The city, in puritanical contrast, was the possible source of anti-communist bourgeoisie sentiment.

Labor mobility was tightly regulated with the *hukou* system of registering every person in specific locations to make them eligible for food, housing, and goods and services. It tried to fix people in place especially in the rural areas. First created in 1958, it made a distinction between urban and rural residents. The system created an urban–rural divide with urban residents having better access. It also effectively limited rural to urban migration. In reality urban enterprises could contract with rural communities for labor of fixed duration. This system is still operating today, more in the breach than in the full observance (Richburg 2010).

There was also the attempt to move people more back to the rural areas. This movement, first begun in 1957, was especially acute during the excesses of the Cultural Revolution from 1966 to 1976 when thirty million urban dwellers were forced to move to and work in the countryside. Shanghai was especially targeted. As a former treaty port, a place of foreign influences and conspicuous consumption, it was treated as a dangerous hybrid place— Chinese, yet contaminated with anti-communist tendencies and bourgeoisie sensibilities. Shanghai with its cosmopolitan past and foreign influences was always viewed with some alarm by the hardliners. Nien Cheng tells a poignant story in her memoir *Life and Death in Shanghai*, of being arrested in 1966 by Red Guards. She was imprisoned for six years, and her actress daughter was murdered. Her "crime" was to be relatively wealthy and to have worked with a foreign company. Prominent local politicians in Shanghai were "purged" in 1967. The city received little infrastructural investment and as a result much of its pre-communist urban fabric remained untouched though renamed. The Bund was renamed Revolutionary Boulevard. The city also boasted an Anti-Revisionist Street and an Anti-Imperialist Street. Shops changed their name by the score to "East is Red."

The *hukou* system and the anti-urban bias of the Cultural Revolution impacted the city. From a peak in 1961 the city's population declined from 6.4 million to 5.4 million in 1978. Over the two decades that elsewhere saw huge urban population gains in developing countries, Shanghai experienced steady decline. Only after 1977 did the population increase.

Under the first phase of communist rule, the years of ideological purity, the city was reimagined around production. Factories were everywhere, both small and large, in the center and peripheral areas. Workers' residential buildings were often adjacent to them. The only other significant new spaces were the theaters for political rallies and mass mobilizations. Squares and mass meeting halls were constructed in the city. People's Square was transformed in 1953 from a racecourse to an open space that could accommodate one million people. The Sino-Russian Friendship Building opened in 1955 covered 80,000 square meters and the Cultural Plaza redesigned from a greyhound racecourse in 1952 was the largest indoor hall in the city. Shanghai became a city of factories, workers residential compounds, and new spaces for political rallies. The city's population was mobilized to work hard, attend rallies and organize themselves for the provision of many public services. The city was restructured so that proletarian production replaced cosmopolitan decadence.

The postcommunist city?

The communist city did not last long after the death of Mao in 1976. After a brief struggle over succession, the party officials purged during the Cultural Revolution re-emerged. By 1978 China's new political leadership was mapping out a policy of pragmatic reform whose ultimate goal was to boost economic growth and raise living standards. State-directed communism had failed. It was time to try something new; a direction that maintained the political power of the CCP but created private incentives and market reforms. Monopoly political power was maintained but decisions were decentralized to the provincial and municipal levels, market reforms were introduced that undermined inefficient state-owned enterprise, and private and foreign investment was encouraged. It was to be socialism with Chinese characteristics, a form of market authoritarianism.

An essential part of the new economic policy shift was to allow foreign investment in the coastal provinces and cities. In 1980 four special economic zones, Shenzen, Xiamen, Shantou, and Zhuhai, were opened up to foreign investment. Four years later fourteen coastal "open" cities were declared. Shanghai was one of the fourteen. After years the city was again to be plugged into global networks and flows. The reglobalization of Shanghai involved, as with other open cities, the creation of special economic and technological zones to foster concentrations in export-led manufacturing, high-tech industries, and financial services.

The national economic policy shift resulted in spectacular economic growth and the consequent massive urbanization of the Chinese population. From 1978 to 2005 the annual growth rate of GDP was close to 10 percent and average wages grew sixfold. The country was transformed from an economic backwater with regular bouts of mass starvation to a major engine of the global economy. The experiment lifted more than 300 million people from absolute poverty and created a large and growing urban middle class. The Chinese economy is now a global player sucking in imports from all over the world and exporting goods across the globe.

Rapid economic growth went hand in hand with accelerating urbanization. In 1980 only about 20 percent of the population lived in cities. In 2008 it was closer to 45 percent, totaling close to 577 million people. By 2015 it is estimated that a majority of the Chinese population will live in cities. The pace of change is striking. Hsing (2009) writes of a great urban transformation as more than eight million people a year moved from the countryside to the city in a twenty-year period. It took the UK 120 years to double its urban population, forty years for the US and thirty years for Japan. It took China a little less than twenty years.

This massive urbanization occurred in the context of newly commodified land and housing market. In 1988 a leasehold land market was introduced. Urban land was effectively commodified with an urban land market trade in land use rights, if not in outright ownership, as urban land is owned outright

Figure 10.2 Shanghai housing: these older neighborhoods in the inner city are fast disappearing in a frenzy of urban renewal.

Photo: John Rennie Short.

by the state. A full and functioning private land market has yet to appear, its full realization hindered by ambiguous rights, political considerations, a dual system of private market and administrative land-use allocation, cronyism, and a general lack of transparency in land dealings. Despite these problems, the virtual commodification of urban land has created commercial opportunities for politicians, municipalities, and private investors. The land market is particularly intense in three areas: in the central city areas where massive urban renewal has replaced and destroyed many old residential and commercial districts; at the central city–inner suburban margins where new residential and economic centers are emerging; and at the rural–urban margins as urbanization continues apace (Figure 10.2). Shanghai has experienced all these types in abundance in the postreform urban landscape of continual change and renewal. The central city areas of open cities have been reimagined from places of political mobilization to spaces for the presentation of a global city. Wu et al. (2007) describe the case of Xintiandi, a 52-hectare site that initially was home to 70,000 people, in the former French Concession area, which now features art galleries, theme restaurants, bars, and high-end shops: "the elegant genre and good designing quality is revealed everywhere, blending with the nostalgia of old Shanghai"(*Streets in Shanghai with Special Features*, tourist brochure, August 2008). The initial plan of 1996 envisioned razing the old housing and replacing it with luxury apartments. One of the buildings housed the site of the first Congress of the CCP. A revised plan

kept this building and some of the other older structures in a mixed develop-
ment project whose marketing slogan was "yesterday meets tomorrow in
Shanghai today." The old dwellings have lost many of their original residents
as the area has gentrified, and as gentrification often dictates, the Shikumen
House Museum now preserves a small corner of the old neighborhood.
Xiantiandi joins Baltimore's Inner Harbor, Sydney's Darling Harbor, and
Boston's Quincy Market as an example of the global urban trend of mixed-
use developments that combine recycled buildings, theme park, shopping mall,
and urban public space.

The most famous inner margin area in Shanghai is Pudong. It was named
Pudong New Area as a new open economic development zone in 1990. Built
on what were rice paddies less than twenty years ago, it is now home to
over three million people. There is a finance zone, a free trade zone, an export-
processing zone, and high-tech park. Its skyline of high-rise new buildings is
one of the most spectacular urban megaprojects that now embody postreform
China, a late modernist urbanism, and a reglobalized Shanghai (Figure 10.3).
The work of big name architectural firms from around the world is on display
in this late modern yet futuristic urban landscape.

The rural–urban fringe is another zone of intense speculative activity. In
the early years of communist rule, the rural fringes were identified primarily
as sites of food production for the city. Now in a postreform era that

Figure 10.3 Shanghai housing: new, high-rise, modernist, residential blocks sprout
up throughout the inner city.

Photo: John Rennie Short.

has witnessed powerful decentralization of people and jobs, they are sites of possible land-use conversion with all the attendant profits to be made from converting rural to urban land use. McGee et al. (2007) write of the "manipulation of the margins" in Shanghai as new residential and industrial complexes are developed along the rural–urban fringe, aided and encouraged by local governments who are key stakeholders in the land development system. The end result is a more widely dispersed urban system with land development often leapfrogging across the landscape as prompted by local governments more than by rational metropolitan land-use allocation.

If one mark of a global city is development-project corruption, then Shanghai has joined the upper echelon. The breakneck urban development, the less-than-transparent planning system and the lucrative profits to be made, all create the perfect setting for corruption and cronyism. It came as little surprise when in 2007 a number of senior officials and business leaders were found to have been involved in shady real-estate projects involving looted government funds, backroom deals, and outright bribery. The people involved were charged not only with bribery but also with leading decadent lives.

The economic reforms also led to the marketization of housing. Prior to 1978, public housing dominated, some of it related to work units. The municipality provided others. The work unit housing was abolished in 1998. In the initial years of reform, public housing was sold off often at very reduced rates. Well-placed officials and functionaries, allocated relatively good housing to begin with, could purchase their housing at much reduced rates. Then the policy shifted again from privatization to encouraging private housing in order to increase the basic supply of housing. The new policy envisioned around 70–80 percent affordable housing, between 10 and 15 percent market priced with between 10 and 15 percent subsidized rental housing, but by the late 1990s market housing became the dominant part of the market, especially in cities like Shanghai. In a commodified housing market differences in wealth are reflected in house purchasing power. And the net effect is a more segregated housing market and a more socially segregated city. Price has escalated in the favored central city areas such as the French Concession where house prices doubled from 2002 to 2005. House prices across the city increased 14 percent in both 2008 and 2009, leading to fears of a housing market bubble.

The socialist transformation of Shanghai was not able to eradicate the precommunist social topography. In the postcommunist private housing market city, the inner areas of the International Settlement and French Concession have retained their status as elite residences. The dominant sociospatial restructuring is gentrification of the inner city and more lower income households pushed to the further peripheries (He and Wu 2007).

In the post reform era the social differences and spatial differentiation have widened. At one end of the continuum are the gated communities of the very wealthy. Pow (2009) describes the emergence of gated communities in Shanghai. These take two forms. First there is the restriction of access to older, upmarket residential areas. The inner ring residential areas of Shanghai are filled with sentry boxes, gates, and the demarcation of exclusive areas (Figure 10.4). There are also the newly built gated communities. Pow describes Vanke Garden City, a 500-acre development built on a former paddy field that now contains a mix of expensive apartments and villas guarded by a 2-meter steel fence and 280 security guards. Outside the gates are chaotic traffic and dirty streets; inside are quiet, well-maintained roads and landscaped gardens. In 1995 such exclusive areas accounted for 12.6 million square meters of residential space in the city, by 2003 they reached almost 22 million square meters. Many of the developments are themed; one development called Olde Shanghais duplicates the Treaty Port Western styles. Others are British-, Australian- or US-inspired with names such as Australia House, Scotland, and Manhattan. The names and the looks of the places are explicitly modern, transnational, and cosmopolitan. It is a modernity embraced in housing developments that "do more than just represent visions and symbols of the good life behind gates but also present, shape and help constitute (middle) class distinctions and subjectivities" (Pow 2009, p. 81).

At the other end of the socio-spatial spectrum of Shanghai are the rural migrants that form what in China is known as the "floating population." These are people registered in rural areas who have moved to the cities for employment opportunities. In 2010 the total floating populating in China was estimated at 211 million, with roughly six million people joining this migrant stream each year. In Shanghai the floating population is estimated at almost five million, or between one-third and one-quarter of the total metropolitan population. As Shanghai has grown, it has continued to attracted migrants from the rural hinterland. In 2000 there were around two million and they constituted approximately 20 percent of the population. In her study of their spatial distribution, Wu (2008) shows how up until the 1990s these migrants concentrated in the central areas, but with the commodification of housing and gentrification of the inner city, they have decentralized to more peripheral areas. These rural migrants face systematic housing discrimination and lack of access to public services of social welfare, education, and health services. Their marginal status is made even more precarious during economic downturns (Cha 2009). Despite their marginalization, they play an important role in the economic globalization of the city especially as a source of cheap labor. Wong et al. (2005) describe the process as one of economic inclusion and social exclusion. The more recent migrants stand out in the streets of Shanghai. They look and sound different, a rural underclass that provide the raw labor power in the transformation of the city.

In the city, as in the country as a whole, income inequalities are widening as the rich and super rich emerge as a distinct economic if not political class. Although ostensibly a socialist country, China has minimal welfare provision to soften the sharpening and hardening inequality of a globalizing capitalist economy. Health insurance is limited, unemployment benefits are rare, pensions are underfunded, and higher education is expensive. The lack of welfare

Figure 10.4 Shanghai housing: a gated community in an older residential area.
Photo: John Rennie Short.

forces many Chinese households into precautionary savings in case of unforeseen downturns.

A middle class is emerging, especially in the bigger and more prosperous
cities such as Shanghai. For this class, incomes increased and new consumption opportunities have opened up. There are even signs of political
mobilization over specific issues. Maureen Fan (2008b), for example, reports

middle-class protests over planned extension of the city maglev train line. This is the dominant form of protest, singular, isolated, and specific to perceived negative externalities more than general resistance to the regime. The naïve belief that growing prosperity will lead to calls for political freedom are perhaps misplaced. As the urban middle class become richer they may be unlikely to promote a democratization that will give power to the impoverished interior, the millions of peasants, and the poorer migrant groups. Rising affluence, while often assumed to lead to greater calls for democratization, can also engender a more pronounced sense of what could be lost in political turmoil and a redistribution of power to the poor masses.

A reglobalized city

The reglobalization of Shanghai involves reinsertion into global circuits of economic and cultural globalization. The economic globalization is most evident in the export-led nature of local businesses and the development of financial services.

There is a global network of stock exchanges. There are also names given to the composite indices that measure the activity in these exchanges. The New York Stock Exchange has the Dow Jones Industrial Average. In London, the index is known as the FTSE. Shanghai now gets a double billing as the name of an increasingly important stock exchange as well as the name of the composite index of stocks and shares traded in this exchange: Shanghai as marketplace and Shanghai as index. The double coding emphasizes the important role of the city in the flow of capital around the world.

The Shanghai Stock Exchange was (re)established in 1990 and is now the sixth largest in the world. The market for securities first began in 1866 and the country's first stock exchange was established in the city in 1891 and by the 1930s it was the financial center of the Far East. It was closed in 1949; its reopening in 1990 in a brand new building inaugurating the beginnings of financial reforms in China and the reglobalization of Shanghai as a major node in the global network of financial flows and transactions. Shanghai has reconnected with global flows of capital and re-emerged as a pivotal point in the geography of global centrality. The city is now an integral part of the consolidation of the global financial industry.

Shanghai's reglobalization is also evident in the self-conscious creation of a modern global city. Since 1991 there has been a frenzy in the remaking of the city. There are the networks of inner ring roads and outer motorways, new subway lines, tunnels and bridges, airport upgrades, and new high-speed trains. There are the new spaces of consumption: the malls, shopping centers, and gated communities. These are brand new buildings that reference an urban modernity. The constructing of new expressways, for example, is not only the building of a new transport system but also marks the enthronement of a fast urban time. The construction of malls, hypermarkets, and hotels is also a direct connection with other global and globalizing cities.

The standardization of these spaces, as these malls and hotels now look exactly like malls and hotels constructed in cities all over the world, is part of their attraction. Serial urbanization signifies a modern city on a par with other major cities across the globe. Even the spaces of government and state agencies in the city are upgraded to reflect the generic corporate chic of company headquarters all over the world. The rich interior, the imposing space and silences, are the affects of understated wealth and uncontested power and authority.

Vast new peripheral developments expand Shanghai's reach far outwards. Shanghai is now a huge polycentric city connected with fast-moving highways, high-speed trains, and contrasting neighborhoods; at the extremes, exclusive gated communities for the wealthy and enclaves of marginalized migrants. The postcommunist city now exhibits marked social and spatial segregation.

There is a fury of urban renewal as large areas of housing are demolished and new developments sprout across the constantly restructured urban landscape. The names given to some of these new developments—Vienna, Cambridge, Garden of Eden—and their transnational forms—gated golf communities, Spanish villas, elegant Victorian town homes, French garden suburbs —are a new space, an urban fantasy space—the hybrid urban modernity of Shanghai.

The city authorities have encouraged this reglobalization. Not only through physical infrastructure and the creation of free trade zones but also through an active place promotion of the city as a place to do international business. Wu (2003) describes the various strategies and shows how global promotion structures local policies such as in the greening of the city to burnish an ecological image internationally. Globalization is not simply ordered by the central authorities or imposed by outside forces but encouraged, managed, and enhanced by the local authority and city elites (Wei and Leung 2005). Wang and Lau (2008), for example, show how the local authorities design and implement strategies for encouraging residential areas for foreigners. In this reversal of Maoist policies, foreign enclaves are promoted and encouraged by local authorities. As the city government shifts from redistributionist– productionist–socialist to entrepreneurial–consumptionist–capitalist, globalization is embraced as the dominant narrative to organize and plan the city. The global city imaginary dominates the understanding and shaping of the city.

The reglobalization of the city also involves a reinsertion into flows of cultural globalization. Gamble (2003) writes of the city's growing cultural re-internationalization from the growth of English language schools to the growing consumption of cultural imports from the West and the increased use of goods and ideas produced outside China. The city is now connected to global flows of people and media images. There is also the reconnection to international culture circuits such as art exhibitions, biennials, festivals, conferences, sporting events, and the like. The largest is the 2010 Expo—see Case Study 10.1. The city is not simply reinserted into existing flows and practices, it is as the case study of the Expo 2010 shows, also reconfiguring and reimagining these events. Expo 2010 Shanghai resuscitates the golden years of the nineteenth-century World's Fairs by its scope, extravagance and ambition, making anew for the twenty-first century an old urban vision. Just as Beijing transformed the meaning of the Summer Olympics, so Shanghai changed the significance of the World's Fair.

Part of the cultural globalization of the city is the increasingly large discursive space it now occupies. One example: in the first 2000 edition of an edited book of international urban readings, *A Companion to the City*, the editors solicited papers about cities from around the world. Case studies included Bangkok, Durban (South Africa), Havana, and Melbourne. There were no chapters on Shanghai. In the second edition of the book, published in 2011, there were three separate chapters on Shanghai. No other city received such consideration in the volume. There was an explosion of interest in the city. It was named the "Most Happening City" by *Time* in 2006. There are the innumerable journalistic accounts; in *The New Yorker* alone for example, Goldberger (2006) deals with architecture in the city while Marx (2008) writes of shopping and consumption spaces. There are also the more academic studies such as Jeffrey Wasserstrom's (2009) perceptive historical analyses of the events in a single year every twenty-five years since 1850 and Jos Gamble's (2003) lively accounts of the effects on the city of flows of money, migrants, and mass media. And there are the many architectural coffee-table books that record the stunning transformation of the city (Pridmore 2008). And no discussion of contemporary urban China is complete without at least one chapter on the city. Campanella's (2008) entertaining and breath-less account of the speed of urban transformation China, for example, has a chapter on the city and the compulsory photograph of Pudong's skyline. What is revealing is not only the sheer number of studies and reports published in just the last decade but the focus, revealed in the titles:

- *Shanghai in Transition* (Gamble 2003);
- *Shanghai: Architecture and Urbanism for Modern China* (Kuan and Rowe 2004);
- *Shanghai: The Architecture of China's Great Urban Center* (Pridmore 2008);
- *Shanghai Rising* (Chen 2009);
- *Shanghai Reflections: Architecture, Urbanism and the Search for an Alternative Modernity* (Gandelsonas 2002);
- *Global Shanghai, 1850–2010* (Wasserstrom 2009).

The terms "global" and "modern" and the sense of change in an upward trajectory recur in different forms and permutations in upbeat accounts of the dizzying rate of change. As one author notes, "Hyperbolic commentaries on Shanghai's reclaimed status as a global city and a great metropolis have become the order of the day" (Wasserstrom 2009, p. 11). But the hype and the enthusiasm are neither misplaced nor undeserving. The city is now an important site for uncovering the complex relationship between urban transformation, globalization, and modernity, an emerging metropolitan modernity that converses with and embraces rather than rejects or distances the past. The old buildings of the Bund were refurbished, the French Concession is now highly prized and there is an active and creative reworking of the themes and aesthetics of the old Shanghai in the reglobalized Shanghai. The modern is a remembered and recreated precommunist history and also a contemporary narrative that is celebrated. The reglobalization of the city involves an excavation of the historical experience before the socialist severance of global ties. Whereas most urban modernities distance and actively forget the past, Shanghai's contemporary narrative chimes with the precommunist past more than recent Maoist experience, which is now marginalized, forgotten, or routinely criticized. The pre-Maoist, more distant past provides the narrative of continuity rather than the more recent Maoist.

The opening week of the 2000 Shanghai Biennale, the first to be opened to international curators and artists, was entitled "A Special Modernity." This special modernity is characterized by a rapidly restructuring urban landscape, the city as both a site for display and a display in itself. The dizzying rate of spatial and social change is at the heart of the experience of urban modernity. From 1988 to 2010 the city grew from 7.3 million to 18 million, the inner city gentrified, new buildings and connections proliferated, housing and land were commodified and speculation was rampant. A speculative capitalism is now the zeitgeist of the city as more citizens actively participate in the stock and housing markets, conspicuous consumption replaces socialist production, and fashion consciousnesses replaces class-consciousness. A culture of money and wealth, display and consumption now characterize the city.

The rapid transformation is a source of some disquiet. The novelist Qui Xiaolong captures the confusion of this modernity when he writes of the speed and jagged directions of change in Shanghai. In his 2010 collection of stories, *Years of Red Dust: Stories of Shanghai*, he tells the stories shared by one group of residents talking around their courtyard late into the evenings from 1949 to 2005. They tell of the communist takeover, the Cultural Revolution, and the more recent breakneck economic growth. As former class enemies become successful businessmen and previous heroes become marginalized, it is a narrative not only of incessant change, but of changing fortunes and the irony and whimsy of fortune. Social categories and political meanings, once firmly established and deeply entrenched, slip and slide and disappear.

A changing urban environment is at the heart of contemporary Shanghai art. In his oil on canvas series of Shanghai, Liang Yunting (b. 1951) records the vanishing revolutionary forms of architecture that are fast disappearing, making the rendition both an act of documentation and an act of memory. A similar process is recorded in the videos of Zhang Jian-Jun (b. 1955), while Yang Yongliang (b. 1980) makes fantasy urban landscapes of skyscrapers and cranes composed to evoke traditional Chinese ink painting (Bessire 2009). Liu Jianhua (b. 1962) makes installations with quotes that celebrate/mock Shanghai's pursuit of global status such as, *Will the first golf course built on a skyscraper's roof appear in Shanghai.* Yang Fudong (b. 1971) uses photographs, films, and videos to capture the experience of the city through the lives of young city dwellers coming of age in one of the most rapidly changing cities in the world. A sense of displacement and longing sits alongside a celebration and embrace of modernity in the work of these young Shanghai artists.

The academic Wang Xiaoming, born in Shanghai, provides a non-fictional account of socio-spatial change in the city. He details the loss of industrial space as former factories close and the urban renewal projects construct malls and new housing. He describes the loss of public political space and its replacement by a commodified space. At one stage he writes, "In a blink of an eye, all that happened has faded from the public memory and been completely erased from the city's architectural space"(Wang 2009, p. 92). His description and sentiment echo Marshall Berman's experience of witnessing the wrecking ball of urban renewal destroying his neighborhood in the Bronx in New York City in the 1950s. Berman's 1982 book, *All That Is Solid Melts into Air*, employs Marx's description of the continual change at the heart of a dynamic capitalism. That the same sentiment experienced in 1950s New York is echoed in early 2000s Shanghai speaks to the global experience of a capitalist urban modernity at its peak of creative destruction.

CASE STUDY 10.1 **Better city, better life**

It cost almost $60 billion, perhaps even exceeding the total cost of the 2008 Beijing Games. It opened on May 1 and ended on October 31, 2010. It hosted almost seventy-three million visitors during its brief but spectacular life. It was Expo 2010 Shanghai.

In Chapter 4 I noted how World's Fairs are an important site in the global network of cities and the transmission of a capitalist modernity. Their high point was in the late nineteenth century, but by the twentieth century, they were eclipsed by the Olympic Games as the dominant global spectacle held in particular cities. Shanghai's hosting of the 2010 Expo elevated the status of the event and the international prominence of the city.

China's capital, Beijing, is the political nerve center of the nation, and it got the 2008 Summer Olympics. Shanghai, as the nation's other major city, sought to host an Expo, of course the biggest one ever. The division of global spectacular spoils fit neatly into the national imaginary of the two cities with Shanghai often depicted as more commercial, more open to the foreign than the capital. Beijing's image is more sedate, and ultimately more boring than the dynamic metropolis now straddling the Huangpu River. The Expo's opening fireworks were purposely not as dramatic as the Olympic opening ceremony in order not to upstage the national capital, but otherwise it was an extravaganza. Images from the Expo are available at www.artdaily.org/index.asp?int_sec=210&p=0&id=454&fid=0.

The Expo was awarded to Shanghai in 2004. From then, giant clocks in the city ticked down until the official opening. An old industrial site on the Pudong side of the river, which previously housed a steel plant, a scrap yard, and around 17,000 people was transformed into the Expo site where pavilions represented every region in China and most countries from around the world. The $5 billion urban makeover resulted in an architectural fantasy land where the pavilion of Spain was designed to look like a wicker basket, the Saudi pavilion featured exotic Arabian gardens, Japan's building was covered in solar cells and France's futuristic pavilion was made from a revolutionary type of concrete. Art works were shipped in from around the world. Denmark sent the Little Mermaid sculpture; Egypt loaned a 3,000-year-old Pharoah's mask; Italy sent a Caravaggio; and France loaned a Gauguin and a Bonnard. The US pavilion was a rather banal structure filled with corporate logos; unlike most other countries, the US pavilion was paid for, not by government, but from corporate support. Furthermore, the US was very late in committing to the Expo, not doing so until Summer 2009.

The main theme was "Better City: Better Life," and the pavilions echoed a consistently green urban theme. Zero-emission vehicles were used, the whole site was powered by solar energy and captured rainwater was the main water source. Secondary themes were a celebrating of cultural diversity and a promotion of economic prosperity in the city.

More than $50 billion was spent upgrading the city's infrastructure to host the event. The promenade along the Bund was rebuilt. New roads and railways were constructed, and an upgrade to the airport reconnected the city to the nation and the world. The city was made ready for foreign visitors. The local population were told to take pride in the city and to avoid spitting, jaywalking, hanging out their washing, and wearing their pajamas in public—a long Shanghai tradition.

When it ended, an area of the city had been redeveloped; a dazzling display of architectural ingenuity was presented; and the ideas of green urbanism were tested, promoted, worked through, and presented. In this global urban spectacle, Shanghai was even more firmly inserted into global flows of people and ideas, and the tradition of World's Fairs/Expos as a site of a transnational modernism was resuscitated and revived.

CASE STUDY 10.2 **Shanghai and Mumbai in comparative perspective**

Cities are sometimes best understood in a comparative framework. Shanghai is one of a number of large globalizing cities in developing countries. Others that spring to mind are Mumbai, Sao Paulo, and Kuala Lumpur. Let us consider Mumbai.

Mumbai has grown dramatically. In 1971 it was already a large city with a population of 7.7 million, but by 2010 it more than doubled its population to reach megacity proportions of nineteen million. It is the largest city in India. Mumbai was a gateway city as India first industrialized and then opened its economy. It is a major port and primary site for textile production and, like many large cities, it has experienced rapid deindustrialization, especially since the large textile strike of 1982, with the proportion of manufacturing employment falling from 36 percent in 1980 to around 25 percent in 2005. It remains India's foremost manufacturing center though it has diversified into entertainment and financial services. There is a large informal sector that has grown in the wake of deindustrialization.

In the wake of the neoliberalization of the Indian economy Mumbai has become more polarized. The range in Mumbai is extreme, from opulence to extreme poverty. Exclusive neighborhoods are encircled by informal settlements; the poor surround the rich. There is, however, a widening segregation as the fortunate few become richer and the mass of the poor become more marginalized. The state caters more to the middle classes than concerns itself with the provision of basic services for the poor. More recently there has been a more active spatial politics of eliminating informal settlements. Harris (2008, p. 2422) writes of the "the new interlinked forms of globalized gentrification and neo-liberal urban policy" that affect Mumbai just as much as London. There is also an urban boosterism among elites that is concerned to make the city as successful in repositioning itself as Shanghai. Yet Shanghai continues to eclipse Mumbai. The rising inequalities in Mumbai lead to a fragmentary and polarized metropolitan space that, as Gandy (2008) notes, is even unable to provide decent drinking water to all its inhabitants. Segbers (2007), in his comparison of the globalizing city project, points to the fact that Mumbai has a highly fragmented governance structure, division between ethnic and linguistic groups, and no explicit and shared goal to become a global city. He refers to Mumbai as "the patchwork city." Shanghai in contrast he labels "The gigantic city inc," with a focused strategy to become a global city, a unitary-party system, a symbiosis between the state and the private sectors, and a driving shared commitment to growth, efficiency, and urban development.

———————————

11 Megalopolis

The liquid city

Megalopolis is a region spanning 600 miles from north of Richmond in Virginia to just north of Portland in Maine and from the shores of the Northern Atlantic to the Appalachians. It includes the metropolitan areas of Washington-Baltimore, Philadelphia, New York, and Boston, covers 52,000 square miles and contains 49 million people. It is the densest urban agglomeration in the US, one of the largest city regions in the world, an important element in the national economy and a vital hub in a globalizing world. Large city regions around the world, like Megalopolis, are the principal hubs of economic and cultural globalization (Figure 11.1).

I employ the term *liquid city* to describe this urban region. Zygmunt Bauman (2005) describes the precarious life lived under conditions of constant uncertainty as *liquid life*. He deploys the term to refer to time and the question of identity in a rapidly changing world. I use *liquid city* with reference to the spatial incoherence of this built environment. Recent growth in Megalopolis as in other urban regions around the world, large and small, has a liquid quality; it is constantly moving over the landscape, here in torrents, there in rivulets, elsewhere in steady drips, but always in the viscous manner of a semi-solid, semi-liquid, half-permanent, yet constantly changing phenomenon. Megalopolis, like many mega-urban regions, possesses an unstable quality that flows over political boundaries, seeps across borders, and transcends tight spatial demarcations; it is a process not a culmination, always in motion, rarely at rest, always in a state of becoming as well as being. Megalopolis is a large liquid metropolis whose boundary demarcation is ever provisional, an approximation, the uncertain fixing of a constantly moving object.

Liquid city grasps the nature of this urban change: it is just the latest in evolving terms for the big urban region. Gottmann (1957) used the term *Megalopolis* in recognition of the spread and transformation of the urban region to a spatially extensive interconnected system of multiple metropolitan areas. Joel Garreau (1991) coined the term 'edge city' with reference to the centrifugal spread of population, office space, and retail. He suggested that new concentrations could be identified at specific edges. Robert Lang's (2003) *Edgeless Cities* challenged this notion and wrote of an elusive metropolis with

Figure 11.1 New York City: the largest city in Megalopolis.
Photo: John Rennie Short.

few sharp edges. And now we have *liquid city*. The term is useful, but it does need to be applied with caution. It does not mean that all is flux. Fixed and frozen elements remain, such as elements of the built form and the persistent patterns of segregation, that are constantly assailed by subsequent change and movement. Large clumps of capital investment such as motorways and airports remain fixed in place, affecting subsequent flows of investment and movement. But flows there are. There are tensions between the fixed nature of the city and urban movements, especially the continual centrifugal forces in Megalopolis. Fixity and flow in a constant dialectic as flows produce new places of fixed investment and concentration that in turn are undermined by new flows. Structure and process, solid and liquid, stasis and flow in a constant interconnected reality. The term *liquid city* highlights the more dynamic element of the dialectic.

Jean Gottmann coined the term *Megalopolis* to describe the urbanized northeast of the US. He first used the term in English in a 1957 article published in *Economic Geographer*, where he prefigured the main arguments that would appear later in the better-known book (Gottmann 1957, 1961). The map that he used in the 1957 paper to delineate Megalopolis was drawn from an earlier study, an innovative government survey to identify economic areas in the country (Bogue 1951). Using data from 1940, as the 1950 Census data had not yet been released, the study used a total of 164 variables (76 agricultural and 88 nonagricultural) for 3,101 counties in the US. A total of 501 state economic areas were identified. The study identified a special class of areas termed *metropolitan state economic areas* where "the non-agricultural economy of such areas is a closely integrated unit and is distinctly different from the economy of the areas which lie outside the orbit or close contact with the metropolis" (Bogue 1951, p. 2). A total of 149 metropolitan state economic areas were identified across the country. A national map of the different economic areas depicted an area of metropolitan economy from Boston to Washington. Gottmann could see the important implications of this map. His genius lay in developing this observation into a wider debate and a deeper analysis.

In the 1961 book, Gottmann describes the region as the hinge of the American economy that contains a concentration of economic and cultural activities, a large number of academic and research institutions, a powerful political center in Washington and a good transportation system that fosters economic connectivity. The book describes a deeply interwoven urban–suburban area with a large concentration of population, supremacy in economics, political and cultural activities, concentrated economic activity, and wealth. The book contains enormous detail, with lots of data, tables, and figures. It is an encyclopedic regional geography in the classic French tradition in which he was schooled: Gottmann completed his doctorate at the University of Paris. But his book is also the recognition and celebration of a new order in the organization of inhabited space, in essence a new way of life. Gottmann saw the region as an incubator of new urbanization, a laboratory for a new experiment in social living. Like many immigrants to the US, Gottmann reveled in the possibility of a society dedicated to the premise of a better tomorrow. He imbibed the new world optimism. The US offered him the possibility of a better future than the war-ravaged Europe he left.

In a careful reading of Gottmann's work, Robert Lake (2003) draws attention to the assumptions behind the analysis: the celebration of size and growth, the sense of inevitability, a belief in unbounded entrepreneurialism and consumer choice in a utopian emphasis on suburban dynamism. Little consideration is given to central city decline or inner city poverty. Gottmann's *Megalopolis* is Manifest Destiny for postwar metropolitan America.

Gottmann's work was enormously influential. Based on the great success of his 1961 book, the term *megalopolis* entered the lexicon of urban studies and entered the English language as the name for any big city region. The study prompted more detailed studies of the northeastern region (Taeuber and Taeuber 1964; Pell 1966; Alexander 1967; Putnam 1975; Borchert 1992). Some were very critical. Weller (1967), for example, found scant evidence of increasing inter-metropolitan division of labor and questioned the validity of the megalopolis as a new community form. Many critics pointed to the considerable gaps in the urban fabric between Washington and Baltimore and derided the idea of a continuous urban region. Peter Hall questioned whether the term was a "convenient fiction, a tool for analysis or has it a deeper function or physical reality?" (Hall 1973, p. 46). Although the concept is still utilized occasionally, it fell out of favor as a tool for analyzing the northeast.

Yet Gottmann's book initiated analyses of large city regions around the world. In a typical work stimulated by Gottmann's writing, Chauncy Harris (1982) identifies the Tokaido megalopolis as the principal urban region of Japan. Such approaches continue today, reinvigorated by the sense that the processes of globalization are most vividly embodied in selected giant urban regions. When Scott (2001) uses the phrase "globalizing city region," he is in effect adopting a more recent version of megalopolis.

The term also became a normative construct, something to avoid through strategic public policies. Osborn and Whittick (1963), for example, argued the case for New Towns as a way to avoid megalopolis. The extensive study of land-use planning in Britain was published in two volumes as *The Containment of Urban England*, with its first volume subtitled *Megalopolis Denied*, to refer to the fact that stringent planning controls averted the fate of the urbanized northeast of the US (Hall 1973). The term continues to be used as a journalistic device in contentious debates about urban growth and land use: examples include the description of the area from San Clemente to Bakersfield in California as a "150-mile megalopolis of overpriced homes on postage-stamp lots dotted with shopping centers and mini-malls" (Anonymous 2003) and the depiction of central Texas in the mid future as "a megalopolis of 2.5 million people" (Schwartz 2003). In current journalistic usage, *megalopolis* is often used to describe the unappealing endpoint of uncontrolled urban sprawl. Debates on the new urbanism and smart growth, for example, are often framed as a way to avoid megalopolis.

Megalopolis today

Gottmann provided the organizing idea, but it is difficult to build directly on his actual definition of Megalopolis since it is neither consistent nor clear. Although the first map in his 1957 paper uses the counties identified in Bogue's 1951 report, subsequent maps in the paper have a different demarcation. I have identified at least six different variants of the region used in his 1961 book. He excluded all of Maine and Vermont when writing about manufacturing, but when writing about agriculture, he included one and sometimes two counties in Maine and four counties in Vermont. Certain counties in New York, Virginia, Pennsylvania, Vermont, Maine, New Hampshire, and West Virginia appear and then disappear as his cartographic representation keeps changing. No definitive list of counties is ever presented in his article or book. And when he employs spatial data, it is never exactly clear what precise definition of the region he is using. Gottmann provided the big idea but no firmly grounded empirical base to build upon.

In this study I use the metric of contiguous metropolitan counties. The US Census identified a Metropolitan Statistical Area (MSA) as a central city with a minimum population of 50,000 and surrounding counties that are functionally linked to the city through levels of commuting, population density, and population growth. From 1950 to 1990, one of the many standard measures was the level of commuting: if a county sent more than 15 percent of its total commuters to the central city, it was considered part of the MSA. In 2000 the threshold was raised to 25 percent. For our purposes, then, the Census Bureau already identifies counties in the orbit of an urban center. Using counties has both advantages and disadvantages. On the one hand they are units of observation that remain consistent over the fifty-year period from 1950 to 2000 and thus allow an easier longitudinal comparison. They are also administrative units with a political reality and planning purpose. They contain the cities of Baltimore and Philadelphia as well as the traditional areas of county jurisdiction. Counties are irregular in size but they provide a consistent frame of reference.

The resultant area of Megalopolis consisted of 124 counties that stretched across 52,310 square miles, twelve states, one district (District of Columbia), thirteen MSAs and the four major metro regions of Boston, New York, Philadelphia, and Washington-Baltimore. There is a close approximation between earlier demarcations of Megalopolis and my own. Bogue's (1951) monograph is a particularly prescient prediction of future developments, a testament to the robust nature of the 1951 government survey. There is also remarkable similarity with the Gottmann demarcation. While there are some minor "disagreements" about counties in the periphery, a function of subsequent expansion for counties excluded by Gottmann, and perhaps an overbounding for some counties included by Gottmann, these occur at the edges, places where the liquidity of the metropolis is difficult to ascertain with pinpoint precision.

Having defined the region, what does it mean? In what sense is Megalopolis a region? It is clearly not a single political jurisdiction, ranging as it does over different states, counties, and municipalities. Neither is it a term in common popular usage. And neither is it a source of identity. People who live in the region rarely refer to themselves as "Megalopolitans." Their identities are linked to a scale well below this giant region. But this disparity between the lived urban experience of individuals on the one hand and the brute existence of a region of contiguous metropolitan counties on the other raises an important issue. The end result of a myriad of individual actions, investment decisions and political choices is the collective, unforeseen creation of a giant urban region larger than our capacity to humanize it and greater than our imagination to conceptualize it. Megalopolis stands as a testimony to the unplanned, collective endpoint of individual everyday decisions. When households move out to the suburbs of Washington, DC, Baltimore, New York, and Boston, the stores that locate along strips leading away from cities are creating a vast urban network that covers the land with a liquid metropolis. We still have a sense that a city should be bounded, comprehensible, with boundaries that we can encompass in our everyday lives and conceptual models. The sheer extent of Megalopolis questions all of these assumptions. The liquid metropolis has oozed beyond our ken. Megalopolis takes us beyond formal city boundaries, everyday lived experiences, and common conceptions of the city, urban growth, and metropolitan frontiers. In a way it is also the product of the provisional, accidental, even surreal quality of the unplanned metropolis.

De Certeau (1984) made a distinction between strategies and tactics. Strategy refers to the spatial ordering of powerful interests; tactics refer to appropriations and transgressions. Through tactics, the strategies of power can be undermined and appropriated. Life in the city is conceptualized as reinforcing both the spatial strategies of the official city, and also the tactical appropriation of the everyday resistances. De Certeau's work is enormously influential in its articulation of the need to connect the official and the every-day, the compliance and resistances embodied in the space–time paths we make across the urban built form and the routes we weave across urban social space. But the strategy/tactics model only takes us so far, and Megalopolis highlights the limits of this conceptualization. Beyond strategy and tactics are unforeseen consequences—new spatial forms of the city that are neither strategic nor tactical, but unplanned, unimagined, and unforeseen. Megalopolis is an expanding universe at the outer limits of the definition of liquid metro-polis, at the very edge of our understanding, and almost beyond our capacity to theorize it.

I will treat the issue of whether or not Megalopolis exists as an epistemo-logical rather than an ontological question. To make sense of this matter, we can draw upon the works of Karl Marx who made a distinction between a *class in itself* and *a class for itself*. A class in itself is shaped by economic and historic conditions, but, if a class for itself is to emerge, there needs to be a conscious sense of shared identity and an appreciation of a shared fate. Similarly we can identify megalopoli that are urban regions in them-selves. However, many things militate against them becoming regions for themselves: political fragmentation, the legacy of separate identities, and the tendency for identities to be more local and national than regional. While economic forces are creating a region in itself, many things operate in the political and cultural realms to stop the creation of regional consciousness. The major cities of Megalopolis, for example, have separate regional news stations and different sports teams. In terms of baseball allegiances, we move north to south through the territories of the Boston Red Sox, the New York Yankees and Mets, Philadelphia Phillies, Baltimore Orioles, and the Wash-ington Nationals. In football terms, traveling south to north is to cross the fierce rivalries between the Redskins, Ravens, Eagles, Jets, and Giants to the New England Patriots. Local school districts, different states and counties, metro TV markets, and fierce sport rivalries all work to suppress the creation of an urban regional consciousness. Megalopolis is a region in itself but not a region for itself.

One example is drawn from the reporting of MacGillis (2006): every working day a group of eight men assemble at around 3.50 a.m. in Luray in Page County, Virginia. By 4.00 a.m. they leave in a bus that takes them 77 miles, north along Route 340 and west on Interstate 66 through the expanding suburbs to George Mason University. The men work in the physical plant shop of the University from 6 a.m. until 2.30 p.m. when they begin their long journey back to Page County. They are not long-distance commuters who have moved out for more space and cheaper houses. They are long-term residents of Luray who have seen economic opportunities in their small town dwindle as the printing plant and the tannery closed. They need employment. George Mason University on the other hand needs workers and can only pay around $33,000 a year, a wage that is half the living wage for a family in Fairfax County. The University uses six vans to ferry in workers from far afield. One of the workers, Sam Dean, has to travel 25 miles just to get to the early morning pickup point; he only has ten hours between getting home from work and leaving for work. Luray is in Page County, which does not even qualify under the contiguous metropolitan county rule that I used to designate counties in Megalopolis. It is off the map yet still linked by complex patterns of commuting. Around Megalopolis the complex patterns of linkages, as exemplified between Page County and George Mason University, connect a dispersed and liquid metropolis.

The mathematician John von Neumann once remarked that we never understand things, we just get used to them. We have gotten used to giant urban regions with their networks that eddy and flow through cities and suburbs across hundreds of miles. Perhaps it is now time to understand them. In this chapter I focus on looking at changes from the baseline of the Gottmann study. I will focus on population changes, economic restructuring, environmental impacts, and the revalorizations within the region. The chapter is a condensed version of a larger study (Short 2007) and more detailed analyses (Vicino et al. 2007). The display of data for such a large area creates difficulties within the confines of a single text. The reader is directed to the electronic atlas of Megalopolis at www.umbc.edu/ges/student_projects/digital_atlas/instructions.htm where more than sixty maps of social and economic data are cartographically presented. The data display and analyses complement and contextualize the discussion presented here.

Population change in Megalopolis

In 1950 Megalopolis had a population of almost thirty-two million people. One in four of all US residents lived in this region. By 2000 the population had increased to over forty-nine million, one in six of the US population. Megalopolis is still the largest single concentration of population in the US. While there are almost eighty people per square mile in the US, there are almost 930 people per square mile in Megalopolis.

In 1950 the population of Megalopolis was concentrated in the urban cores: over one in every two of the total population lived in the central cities. By 2000 much of the population and vitality of the region had shifted to the suburban counties. The region changed from a big city population to a much more fully suburbanized agglomeration. Two out of every three people now live in suburban counties. It is important to note that the traditional "city–suburb divide" no longer suffices as a standard measure of comparison. Puentes and Warren (2006) identify what they call "first suburbs," defined as counties that were metropolitan counties adjacent to a metro core in 1950. These inner ring suburbs grew very quickly from 1950 to 1970, leveling off after 1990. Since the 1980s many of these suburbs have become suburbs in crisis as they lose population and experience declines in tax base and house prices (Hanlon 2009). The devalorization of these inner ring suburbs is the defining characteristic of what has been termed "suburban gothic" (Short et al. 2007).

While the fifty-year data range shows decline in central city areas, since 1990 there has been a small but significant rebound in certain cities. Between 1990 and 2000 central city areas in New York such as the Bronx and Queens increased their populations by respectively 10.7 percent and 14.2 percent, representing absolute increases of 128,861 and 277,781. Boston increased its population from 562,994 in 1980 to 590,763 in 2006. Because of their growth in producer services and immigration flows, New York and Boston are experiencing a small population growth. Cities such as Philadelphia and Baltimore with less buoyant economies saw decline. In both cities the remaining population was proportionately more poor and black.

The racial mix of the Megalopolis population has changed. Whites have redistributed from the central cities to the suburbs. In 1960 83.8 percent of the population of the central cities was white, but by 2000 this had declined to 42.4 percent. The whites became less urban and more suburban. White flight left proportionately more blacks in the central cities and especially in the cities experiencing greatest economic difficulties. Blacks have doubled in population from 1960 to 2000 and now constitute 16.8 percent of the region's population. In 1960 blacks constituted 15.7 percent of the central city population, by 2000 this figure had increased to 27.4 percent. They have become the majority population especially in cities with significant job and population loss. But there is also black suburbanization. In 1960 there were 773,160 blacks in the suburbs of Megalopolis but by 2000 this number had climbed to almost three million. Prince George's in Maryland, for example, a suburban county that borders Washington, DC has a majority black population. It is home to over half a million blacks, who constitute two out of every three people in the county.

The Asian population has increased in numbers from a relatively insignificant 87,000 to over 2.3 million. Asians are in the central cities as well as the suburbs. There are now almost six million Hispanics in Megalopolis constituting 12 percent of the total population; three-fifths are in the central cities, the remainder in the suburbs. The growth is particularly marked in the major metro areas of New York, Washington, DC, and Boston. In selected cities and counties, the Hispanic population is the major driver of demographic growth.

Environmental impacts

The US population leaves a heavy footprint on the earth. In Megalopolis there are now nearly fifty million people living at relatively low densities and consuming large amounts of energy. Compared to 1950 there are more people driving more cars to more places; more people running dishwashers, flushing toilets and showers; more people in more and ever bigger houses. Whatever the measure used—automobile usage, water consumption, or waste generation—it is a similar story of increasing population growth in association with increased affluence and spiraling consumption, producing a greater environmental footprint and increased strain on the natural systems that sustain and nurture life. As more population crams into the region, an incredible environmental transformation is wrought. Close to fifty million people, with the greatest environmental impact per head in the history of the world, now live in Megalopolis.

Economic restructuring in Megalopolis

Over the past fifty years, a major economic change has been the decline of manufacturing, the growth of services, and the growing importance of government. Manufacturing has long played an important role in the life of the region as a significant employer and major source of revenue. In 1900 Megalopolis had almost one in two of all manufacturing workers in the entire country. By 1950 this number had fallen to one in three. By 1997, the numbers had fallen to 1,498,706, only 12.3 percent of the national total. There was a significant deindustrialization of the region in both absolute and relative terms. The region has lost over 1.5 million manufacturing jobs since 1958 and is no longer the manufacturing powerhouse of the US economy.

The region retains its primacy in selected producer services. Over one in every two workers in the nation in the important sector of finance and insurance workers are located in Megalopolis, with one in every ten workers based in the New York metro area. The figures are even higher for the sub-category of Securities Intermediation—81 percent of all workers in the US in this category are employed in Megalopolis, with 33 percent located in the New York metro area. Although Megalopolis shed its manufacturing jobs, it is home to information processing sectors. Megalopolis contains 55 percent of all workers in the category of Professional, Scientific, and Technical. It is the analysis of information rather than the manipulation of metal that is now the defining economic characteristic and leading economic sector in Megalopolis.

Economic growth is heavily dependent on the role of government. Not only at the local and state level but especially at the federal level. Government spending influences private market decisions. The location of public highways, for example, has guided the form and level of private investment in suburban areas. The edge cities of out-of-town shopping malls and bedroom communities are as much creations of public spending as they are functions of private investment. Public investment provides an important container for private investment. Government spending also plays a role in the location of fixed-asset investments such as military bases and research centers. One of the fastest-growing counties in Megalopolis is Montgomery County in Maryland. Its population grew from 164,401 in 1950 to 950,680 in 2008—a 578 percent increase while the increase for Megalopolis as a whole was only 53 percent. The county has a concentration of federal research laboratories and regulatory agencies that in turn attract high technology companies, service industries, and vendors. Montgomery County is home to nineteen major federal research and development and regulatory agencies, including the National Institute of Standards and Technology, the National Institutes of Health, National Oceanic and Atmospheric Administration, Naval Medical Center, Nuclear Regulatory Commission, the Food and Drug Administration, the Department of Energy, Walter Reed Army Medical Center, U.S. Army Diamond Labs, and the Consumer Products Safety Commission. The National Institute of Health in Bethesda, for example, houses twelve research institutes employing 20,000 workers and a budget of $28 billion. The National Institute of Standards and Technology employs 2,600 scientists at its primary site at Gaithersburg in Montgomery, developing measurement standards necessary for industry commercialization. The Food and Drug Administration (FDA), headquartered in Rockville in Montgomery employs 4,500 people. With this steady injection of federal dollars and the creation of secure and well-paid employment in the scientific research sector, it comes as no surprise that Montgomery ranks as the ninth most affluent county in Megalopolis with a median household income in 2007 of $92,440; the average for the nation was $50,740.

In the past fifty years Megalopolis has undergone a profound economic transformation that includes a decline in the amount of land devoted to agriculture, a marked loss of manufacturing employment, the growth of services, the rise of government as a powerful economic motor, the suburbanization of retail and the overall shift of jobs from cities to suburbs. Each of these trends have distinct redistributional consequences including the decline of blue collar jobs and a weakening of organized labor, the rise of female employment participation rates, and the restriction of job opportunities for those trapped in the inner city by limited mobility as more jobs suburbanize.

The revalorization of Megalopolis

What underlies and embodies these population and economic trends is a profound revalorization of metropolitan space. Capitalism is a system always in motion. In their 1848 *Communist Manifesto*, Marx and Engels referred to the "Constant revolutionizing of production, uninterrupted disturbance of all social conditions, everlasting uncertainty and agitation . . . All that is solid

Figure 11.2 Abandonment in inner city Baltimore.
Photo: John Rennie Short.

melts into air." Almost a hundred years later, in 1942, Joseph Schumpeter referred to the creative destruction at the heart of the capitalist system. Megalopolis is both a primary container and important vehicle for such dynamism. The immediate post-World War II era saw massive public and private investment in the suburban fringes. Capital reinvestment was fixed into the suburban landscapes in the form of houses, roads, factories, stores, and infrastructure. But, since the 1970s, there is a revalorization that has re-made Megalopolis into a new metropolitan form marked by complex patterns of growth and decline, expansion and contraction. From 1950 to around the mid 1970s, the primary dynamic of the US metropolis was a suburban shift. Since then, the picture has become more variable, with at least four investment/disinvestment waves.

First, there has been a reinvestment in the central city. Downtown business interests responded to the postwar suburbanization of business and customers by initially promoting and supporting urban renewal programs as a way to maintain the commercial viability of downtown. Urban renewal programs of the 1950s and 1960s were attempts to stem the tide of decentralization and preserve downtown property values. The attempted solution not only failed, it exacerbated decline as downtowns became filled with unattractive, sterile, unusable spaces and a depressing collection of dead zones beneath elevated highways and busy intersections. A new strategy emerged in the 1970s and 1980s that focused more on building than demolition, on entertainment rather than production, and on public–private partnerships rather than a reliance on only either private risk-taking or on federal programs. Alliances of civic leaders, investors, and developers in cities across the nation sought to construct a new urban iconography of themed retail districts, cultural centers, conventions centers, stadiums, and festival malls, all in a reimagined downtown. The goal was to halt the devaluation of the downtown through its promotion as a place of fun, frivolity, shopping, and spectacle. The new downtown was imagined as a festival setting and was promoted as a cultural centrality in a splintering metropolis. In order to secure this new writing of the downtown, public money had to underwrite private projects and citizens had to be convinced that the benefits would ripple through the rest of the urban economy. A classic example is Baltimore's Inner Harbor. By the 1950s, much of the old port, right in the heart of the city, had become an abandoned space (Figure 11.2). A cluster of developments built between 1977 and 1981 made the Inner Harbor a festival setting: the World Trade Center and Maryland Science Center were built with state and federal funding; the Convention Center was funded with $35 million from state funds; Harborplace, built by the Rouse Corporation, provided retail and restaurants in two large pavilions. The National Aquarium was built with $21 million from the city council. The Hyatt Hotel provided a downtown anchor after a $12 million public grant was made to Hyatt. There was a public underwriting of the whole redevelopment.

A second trend, and very much related to the first, has been the return of people and capital to selected parts of the city, often given the general name of "gentrification." The original definition of gentrification referred to housing that passed from lower- to higher-income households. It has now taken on a number of different meanings to include the general sense of displacement of lower income households and the arrival of new single-person and non-child households, the yuppies of popular discourse, either into new or refurbished dwellings (Figure 11.3).

The move back to the city has been explained by a number of factors, including, on the demand side, the persistence of high income jobs in central city locations, new forms of household formation with smaller households, more single person households and non-child households who place less emphasis on city school systems, and more emphasis on accessibility to employment and urban recreation. On the supply side, Neil Smith (1996) identifies an investment opportunity created by the difference between the land value and the potential land value of an accessible inner city location given the changes in demand just outlined. This "rent gap," as he terms the difference between existing and potential value, creates new opportunities for investment. In a detailed study of New York's Lower East Side, he identifies turning points in the housing investment change that occurred slowly, first around 1976, and reaching its greatest extent in 1979–80. Gentrification often comes with displacement. Newman and Wyly (2006), for example, found that between 8,300 and 11,600 households are displaced each year in New York City. There was also reinvestment in the city that was unconnected to gentrification. Wyly and Hamel (2004) report that redlining, the practice of not lending to

Figure 11.3 Gentrification in inner city Baltimore.
Photo: John Rennie Short.

inner city neighborhoods, was replaced by subprime mortgage lending to inner city residents. This was an important factor in the subprime collapse and subsequent economic meltdown of late 2008.

New York is home to a number of gentrifying projects. Major construction projects such as Towers on the Park, which opened in 1988, and individual households buying up brownstone dwellings led to a distinct form of gentrification along the western corridor of central Harlem. In 1994 Harlem became a Federal Empowerment Zone that enabled tax credits, and federal and state monies. A large indoor mall opened in 2000 with chains such as Disney, Old Navy, and HMV that previously ignored Harlem.

Just across the Hudson River from Lower Manhattan sits the small, formerly working-class city of Hoboken, just a quick ferry ride across from New York City's financial district. The process of gentrification began around 1980. Controls that maintained the presence of lower income households such as tenant protection and rent control became more lax. The housing stock was effectively emptied of lower-income residents in favor of the higher-income households. As the bull market of the 1990s raised the number and remuneration of financial service workers in Manhattan, the process of gentrification accelerated in Hoboken and extended into such places as Newark and Jersey City.

Broad cycles of gentrification coincide with the cycles of the market, especially the rise and fall of business services that require downtown locations. The recession of the early 1990s limited gentrification in selected cities, the economic upturn of the late 1990s produced another cycle of gentrification, while the downturn of 2008 will no doubt see a curtailment of speculative growth.

Third, there is an effective disinvestment from the inner ring of selected working class and middle class suburban neighborhoods as the demand for many of these neighborhoods shrinks. Many of the small, single-family house suburbs that were constructed and grew in numbers from 1950 to 1970 have been devalorized. The suburbs in crisis are most prevalent in metro areas where postindustrial contraction was not offset by an increase in well-paying service employment.

Fourth, the suburbs are also places of the wealthy, where new rounds of housing investment, brought about by the increasing wealth of the already wealthy, reinforce old established landscapes of privilege, as well as creating new landscapes of consumption and exclusion. Affluent suburbs are home to the established wealthy as well as the new wealthy, to the understated rich as well as the conspicuously rich.

The net effect of these changes is suburban places with very divergent experiences. Table 11.1 lists the five types of suburban places identified by principal components analysis and consequent grouping of component scores (Vicino et al. 2007): affluent, underclass, black middle class, immigrant gateways as well as the traditional suburbs, often termed "Middle America." The affluent category included such places as Scarsdale, New York, which

Table 11.1 Suburban places in Megalopolis

Demographics	Income	Education and employment	Housing	Examples
Affluent places Mostly white; married parents	Very high income; low poverty	College graduates; management occupations	Newer, large housing stock; high homeownership rates	Scarsdale, NY; Chevy Chase, MD
Underclass places Black; Hispanic; single-parent families	High poverty; low income	High school drop-outs	High rental; older housing stock	Camden, PA; Asbury Park, NJ
Black middle class places Significant black population; some single-parent families	Middle income; low poverty	College graduates; high public sector employment	Built after 1970s; high homeownership rates	Bowie, MD; Mitchellville, MD
Immigrant gateway places A quarter foreign born; Hispanic and other races high; mostly married couples with children	Low to middle income; some poverty	College graduates; some high school drop-outs; varied education levels	High rental; low homeownership rates	Hoboken, NJ; Tysons Corner, VA
Middle America places Mostly white; married families; "1950s image" of suburbia	Low to middle income; low poverty	High school graduates; some college	Mostly homeowners; postwar bedroom communities	Levittown, NY; Dundalk, MD

Source: after Vicino et al., 2007.

had a median household income of $182,792 in 2000, almost four and a half times the median household income of the New York metropolitan area. The population of Scarsdale is highly educated with 80 percent of the population college graduates. Many members of the workforce in Scarsdale are managers and professionals with 20 percent employed in finance, information, and real estate (FIRE), and 26 percent employed in the health and educational fields.

Camden, New Jersey, is an example of an urban place that experienced tremendous industrial growth in the early twentieth century and subsequent marked decline. The gramophone was invented in the Nipper Building, home of the Victor Talking Machine Company, later the Radio Corporation of America (RCA). At its peak in the 1950s, RCA employed over 20,000 people in the city spread over fifty buildings. By the 1960s the decline set in as manufacturing employment disappeared. The city lost jobs and population. Retail, and the middle class have moved out to suburban New Jersey. From a total of 125,000 in 1950, this city now has a population of almost 80,000 people. Almost half of the population is black, and 40 percent is Hispanic. The poverty rate increased from 20 percent in 1970 to 35 percent in 2000 and the median household income was $23,421 in 2000, less than half the median household income of the metropolitan area. Almost half the population did not graduate from high school.

Megalopolis is home to significant amounts of immigration from overseas. The foreign born population has increased from 10 percent in 1960 to 20 percent in 2000, almost double the national average. Migrants are found in both central cities and in suburban areas; particular concentrations can be identified as immigrant gateways. One example is Tyson's Corner, VA, an archetypal edge city located off the Washington, DC Beltway. The population of Tyson's Corner is 18,540 with almost 35 percent foreign-born, and 70 percent have four-year university degrees.

There have also been significant changes within the central cities. Mass suburbanization has siphoned off the middle-income white families from the city. In more recent years there also has been a suburbanization of the black middle class. In the wake of this movement, cities have become very polarized as increasingly only the very poor and the very affluent remain. In their study of household income distribution in US cities, Berube and Tiffany (2005) show that the largest cities in the nation house a disproportionate share of low-income households. Using their typology of the 100 largest cities in the country, we note that cities in Megalopolis fall into three main categories:

- *Stressed*: where there are twice as many households in the bottom two quintile of income distribution as in the top two; cities in this group include Baltimore, Philadelphia and Newark.
- *Low–moderate*: where there are more households in the poorer categories than in the richer categories, although not as marked as in the stressed category; examples are Jersey City, Boston, and New York.

• *Divided*: where there is a U-shaped income pattern with very few middle-income households; cities in this category include Washington, DC, where the median family income range by census tract extends from a high of $345,117 to a dismal low of $17,592.

The selective nature of suburbanization has left behind the affluent and the poor. Often deep racial cleavages underlie these wide income differences.

This briefest of snapshots indicates the growing diversity of Megalopolis. As inequalities have widened, the difference between rich and poor places has increased. The suburbs, for example, are no longer just the preserve of the middle class; they contain rich and poor, black as well as white. And the increased levels of immigration have also created clusters of immigration gateways. Compared to the more homogeneous suburbs of the 1950s, the revalorization of residential space in association with increased immigration and growing inequality, creates a more complex residential mosaic.

Megalopolis as political entity

We have already related the Marxist distinction between a social class in itself and for itself to similar distinctions at work in massively liquid metropolitan regions. As housing and job markets extend out from one metro area to another in an increasingly interconnected network, many things operate in the political and cultural realms to inhibit the creation of the wider regional consciousnesses. Similarly, local school districts, different states and counties, metro TV markets and fierce sport rivalries all work to suppress the creation of a regional identity.

Megalopolis is one of the most important urban regions in the US and indeed in the world. Its internal coherence has deepened as lengthening journeys to work, widening regional job markets, and dispersing housing markets effectively link the separate metros into overlapping fields of influence and interconnecting flows of people and goods. The region is *in itself*. Whether it will become a region *for itself* is a more debatable issue. At the moment, identities, allegiance, and political realities all work to balkanize Megalopolis.

There are some—public choice theorists being the most vocal—who would argue political fragmentation is a healthy state of affairs. A large number of different municipalities allow residents to choose a variety of tax loads, school districts, and forms of government. However, there are also problems associated with this fragmentation. We can consider two.

The first is a central city–suburban fiscal disparity. Central cities, especially those with shrinking populations, have a declining tax base and lower income population while the suburbs have an expanded tax base and a relatively affluent population. Cities rely on the property tax as a source of revenue. With declining population and an out movement of businesses and higher-income households, the tax base shrinks while the concentration of poorer

Figure 11.4 Low density commercial development in Megalopolis.
Photo: John Rennie Short.

people places greater demand on service provision such as police, welfare, and social services. The older cities also have an ageing infrastructure that is expensive to maintain and replace. Central cities have to deal with the politics of economic decline while many suburbs contend with the management of growth. Municipal fragmentation that separates out poor cities from affluent suburbs reinforces the inequalities in US society (Figure 11.4).

A second related problem is public education. In the US, the federal government has a very limited role in providing funding and resources to public schools. School funding is dominated by state and municipal sources. States provide on average 50 percent of total school budgets, local districts around 45 percent, based on local property taxes, while the federal government only constitutes 5 percent. At the school district level, disparities in wealth feed directly into educational standards and performance. Poor school districts cannot afford to spend the same as richer school districts. The result is a range of school districts, some, more often in the affluent suburbs, are platforms for success, while others, especially in the cash strapped central cities, are funnels of failure. The political fragmentation of the region into a large number of local governments has profound social outcomes.

Megalopolis revisited

Gottmann's original Megalopolis, based on analysis of 1950 data, was the manufacturing hub of the national economy, with substantial central city populations, few foreign-born, and marked racial-ethnic segregation. The most significant changes in the intervening years include the relative shift of population from the central cities to the suburban counties, the loss of manufacturing jobs, especially in the central cities, the growth of services, and the increase in the foreign-born population, especially in the selected central cities, although suburban counties also witnessed the absolute and relative increase in the foreign-born. Despite all these changes, racial segregation remains a stubborn fact of life in the nation's largest urban region.

A liquid city has wider cultural implications. Bauman (1992) also writes of a postmodern condition marked by a loss of certainty about conduct, the unpredictability of change, a lack of centeredness, the decline of grand narratives. The recent evolution of Megalopolis embodies this shift. As rapid changes restructure our cities, as more of our lives take place in a decentered metropolis, as city centers decline, as we follow space–time paths through the metropolis seemingly disconnected from our fellow citizens, and as our truly public spaces shrivel to a series of highly segmented places, then the oppressive and unsettling sense of continual change is made very real, the loss of grand narratives is made palpable, and the lack of center to social life is made visible. The postmodern condition thrives in Megalopolis.

The Scottish polymath, urban scholar and planner Patrick Geddes (1854–1932) first employed the term *megalopolis* in 1927. For Geddes it characterizes a degenerative stage of urban development in an era of giant cities, after the vitality of *metropolis* and before the exploitative *tyrannopolis* and the finality of *necropolis*, the city of war, famine, and abandonment. In 1961 Gottmann, in contrast, saw Megalopolis, the place, as a harbinger of a new way of life. Which is correct, the gloomy formulation of Geddes or the sunnier claim of Gottmann? The answer rests on whether we can build livable humane cities that provide employment, hope, and sustainable futures. Megalopolis poses problems, major problems no doubt, but the forward edge of history lies in the collective and shared fate of such giant city regions as Megalopolis.

CASE STUDY 11.1 **Why here?**

This chapter emerges from work motivated by three reasons. First, I moved to the Baltimore-Washington area in 2002. I remembered reading Gottmann's *Megalopolis* as a student and realized I was now living in this urban region. Situated between Baltimore and Washington, frequently traveling along I-95 and regularly taking the train to New York, I was acutely aware of the spatial spread of the region. Studying where I live has been a feature of my research going back to my graduate days at the University of Bristol when I studied the dynamics of the local housing market. It continued when I moved closer to London and studied the M4 growth corridor that stretched from London to Reading. Later, when I moved to the US, I examined the renewal and reimagining of Syracuse NY as it tried to move beyond its rustbelt past and industrial image. Megalopolis was another opportunity for a locally based study.

Second, because of the importance of Gottmann's work I wanted to do an update prompted by the simplest of questions: What has happened in the region since the book was published? We have so few good baselines studies in the social sciences that it was a rare chance to do a follow-up study to ascertain elements of change as well as dimensions of stability.

Third, a study of the region provided an opportunity to do a case study of a giant urban region. Peter Taylor (2004) describes the world economy as structured around an archipelago of global city regions. In the developed world these city regions are the loci of control and command functions with a heavy concentration of advanced producer services such as banking, advertising, and business services. In the developing world, these are the site of multinational corporation investments and new techniques of manufacturing as well as centers of service industries. To study Megalopolis is to undertake a case study of one of the larger and more important mega-urban regions in the US and the world, allowing an examination of one of the major building blocks of the globalizing economy. This study of Megalopolis represents a Janus-like sweep, considering both what has happened to one of the worlds' largest and most important urban regions and the likely course of future developments.

The work was prompted by a personal quest, a backward glance at a classic study, as well an assessment of an urban form that is beginning to dominate both the national and the global economy.

CASE STUDY 11.2 **Living in the liminal**

I inhabit a liminal space in the vast sea of suburbia in Megalopolis. I live midway between Washington, DC and Baltimore. My job is closer to Baltimore but I read the *Washington Post* rather than the *Baltimore Sun* and my local television news comes from DC not Baltimore. I live betwixt and between the two urban poles of the Washington-Baltimore metro region.

Sprawl and urban spread have covered the spaces between urban centers especially in places like Megalopolis. Often there is no sharp distinction where one place ends and another begins. The southern Baltimore suburbs merge into the northern DC suburbs. And only slowly does the *Post* replace the *Sun* or the support for the Baltimore Ravens and Baltimore Orioles replace the signs of support for the Washington Redskins and Washington Nationals. It is a place where the edges form a whole, as encroaching suburbs transform out of town shopping centers to the middle of the urban region shopping centers.

I live in this liminal space where the unfolding frontier of suburban expansion is a place of edges. I live in a region where edge land uses, places such as Laurel Race Track, BWI airport, the National Security Agency, the military base of Fort Meade, nature reserves and military bases, wildlife refuges and industrial parks, green spaces and WalMarts, make up a complex and often confusing urban fabric as they are incorporated into a chaotic mosaic. This is not the homogeneous banality of the suburban image but the diverse and complex reality of metropolitan suburbs that include the immigrant gateway, affluent enclaves, pockets of poverty most clearly embodied in the trailer parks, the old suburbs in crisis, as well as the middle-income, middle-class suburbs of the popular suburban mythology. We are only just beginning to identify, let alone describe or theorize, this metroburbia of edgeless cities (Figure 11.5).

To live in liminal space is to face discursive uncertainty. "Where do you live?" is a common and an easy question to answer on the surface, but more difficult to answer if you live in the metroburbs. People want single statements. To answer that I live in a town called Laurel that is part of metroburbia between Baltimore and DC, is accurate but not socially smooth. And if one-word easily recognizable names are required, do I say Baltimore or Washington? Do I want to impress them with my access to a political power center or make a claim to a certain urban grittiness?

I live in a liminal space of discursive silences. The television and radio traffic reports, for example, are centered on the two main cities not on those who live on the dual edge. My particular traffic experiences, like so much, are not given accurate representation.

Figure 11.5 The ever expanding edge of a liquid city: suburban growth in Maryland.
Photo: John Rennie Short.

What time is this place? Kevin Lynch (1972) drew attention to the urban spacing of time and urban timing of space. The dominant timing of my everyday mobility is the rhythm of traffic. Living in a place where driving a car is a necessity, I have to manage my day around the morning surge and the early evening swell of the daily auto rush hours.

What sound is this place? Graham Gilloch (2007) has resuscitated this question in his discussion of the critical theorist Siegfried Kracauer. As befits a liminal space, the remembered sounds of my neighborhood are varied. There is the dawn chorus of birdsong, the deadening crickets in the hot summer nights, and the occasional hoot of an owl on a quiet winter's night. There is also the sound of sustained gunfire: the Secret Service firing range is well within earshot. There are the occasional sounds of chopper blades as officials fly immediately overhead from Washington into and out of the nearby NSA. There are also the silences. The quietness of the night is something I especially treasure after visits away to noisier cities and towns.

There is a connection between the time of this place and the sounds of this place. As someone who works a lot from home I am very sensitive to the daily noises in my residential area. I live in the suburbs; low density with lots of greenery, and that is the problem. All that greenery has to be whacked and manicured in order for the suburban image to be achieved and maintained. Grass has to be cut in summer, leaves blown in fall, snow plowed in winter, trees and bushes pruned in both spring and fall. None of this is done slowly and quietly by hand. No. Brute, electronic force does it all. My neighbors, each with the carbon footprint of a medium-sized African city and the mechanical clout of a small European army, wield an ear-wracking array of power tools. They use these on the weekend. During the week, the hired gardening crews marshal their own arsenals of landscaping weaponry whose decibel levels resemble sticking your head three feet from the jet engines of a 747. The end result is a neighborhood filled with the electrified screech and aching whine of the machine.

In the suburbs of Megalopolis, as in many other giant urban regions across the world, peripheries become nodes and the marginal turns into a new centrality as edges are incorporated into the smooth surface of new metropolitan arrangements. The narrative uncertainty of a single-centered city is replaced by the uncertainty and ambiguity of metroburbia. The decline of grand narratives and associated uncertainty and ambiguity that marks late modernity/postmodernity is made palpable and given embodiment in the liminal suburbs of metroburbia.

Part 4

Urban utopias

12 Postscript

For the first time in human history more people live in cities than in the countryside. The urban experience is now the dominant human condition, so it is perhaps time that we treat cities seriously as the environment of the majority of humans. For too long urbanization has been portrayed as a fall from grace, a move from the natural to the unnatural, from arcadia to dystopia. As someone who lived in a small village and worked for some time on farms, I have no such illusions about the pastoral life. Over two hundred years ago the poet Robert Burns described the rural laborer's work life as "the cheerless gloom of a hermit with the unceasing toil of a galley-slave."

The shift from rural to urban is quick and dramatic. While the reality of the world around has changed quickly, ideas we have of the world are often slower to respond. Many of our normative ideas about city life derive from a pastoral existence or a rural world. It is time we view cities as arenas for interaction, rational economic contexts, and sources of innovation rather than dismiss them as sores upon the body politic, economically irrational and socially regressive. Urban growth has persisted for two centuries because there are powerful economic and social–political forces at work. Cities have grown because people have voted with their feet. Cities are places of opportunity and creativity; they are the future.

There are powerful economic forces behind urban growth. Cities allow the efficient clustering of firms and workers. Large cities provide a "thick" market (a large pool of labor and specialized firms), market access and savings in the provision of public goods. In effect, cities, because they concentrate economic activities, make markets more efficient. To be sure, there are negative externalities. The high costs of congestion, pollution, and the large pools of unemployed and underemployed also mark city life. Yet the spatial concentration of cities creates a favorable economic environment by providing the technological spillovers and socio-cultural networks that support continued concentration. There are efficiency gains from the scale effects of urban centers. Cities exist because, simply put, they make markets more efficient. They allow the easy transfer of information and ideas. The geographic proximity in large dense cities facilitates face-to-face contact and the maintenance of trust. Cities, despite the side effects and costs, maximize market productivity.

Around the world people keep moving to cities because that is where there are the greatest economic opportunities and better access to public goods and services such as education and health. When given the choice and the opportunity, many of the young and the ambitious make their way to the large cities of the world.

The economic rationale for cities is easy to muster. But what general ideas should we draw upon if we want to consider a fairer and more socially just form of urban living? What ideas can we draw upon to foster and sustain livable, humane cities? Let me briefly consider three sets of ideas that cluster around both the paradoxes and the opportunities of urban living.

Collective security and individual freedoms

We live in fluid, fast-changing times. Cities are the stage for economic re-structuring, social change, and political cleavages. In the wake of 9/11 and the economic meltdown of 2008, the sense of danger and uncertainty has increased. Collective security takes many forms from providing collective support to ensuring the physical safety of citizens.

Let us first consider security in an everyday sense of adequate provision of public goods and services. As economies grow a basic social cohesion requires more collective provision of goods and services such as health and education. There are obvious constraints. Needs are always infinite and resources always limited. Health spending, for example, can soon spiral ever upwards without some form of strict cost controls. As countries become richer the population tends to age putting more pressure on a declining number of productive workers to finance the needs of an ever increasing and ever more demanding group of elderly people. Generational inequities are building in much of the developed world—see Case Study 12.1.

In the cities of the developing world there is the problem of basic poverty: over 3 billion people live on less than $2.50 a day. Inadequate public goods and services reinforce this poverty. Over 400 million people have no access to safe water and 270 million have no access to health care. Almost 25,000 children under the age of five die every day.

There are no easy solutions. However, we need to set standards and goals for improving the lives of ordinary people, rather than pursuing abstract economic growth statistics such as increasing GDP. The United Nations has established general principles with its Millennium Goals (see Table 12.1). Those concerned with global poverty often favor aid transfers as a way for rich countries to help poor countries. Others suggest a more targeted approach with a greater role afforded to microfinance and the encouragement of tar-geted direct foreign investment. Too much aid is captured by the elites rather than percolating down to ordinary individuals. Government to government aid is often not very effective.

The story has some bright spots. There has been some progress. The number of people in extreme poverty has declined from half of the developing

Table 12.1 United Nations Millennium Development Goals

1	Eradicate extreme poverty and hunger
2	Achieve universal primary education
3	Promote gender equality and empower women
4	Reduce child mortality
5	Improve maternal health
6	Combat HIV/Aids, malaria etc.
7	Ensure environmental sustainability
8	Develop global partnership for development

world population in 1990 to a quarter in 2005. However, much more needs to be done. Opening up rich country markets to producers and suppliers in developing countries is an important task. And when working out the details, that requires a greater commitment to fair trade as opposed to the mindless enunciation of free trade. I also suggest much more direct involvement by people of the richer world through microfinance, community-linked initiatives and city-to-city contacts. We have a shared future that we need to reclaim from finance capital, state bureaucracies, and multinational nongovernmental agencies. I am enormously hopeful that with the Internet and cultural globalization we have a better grasp of what is happening in places around the world. There is a science fiction trope of one global city. It is recognition of the possibilities and necessary reimaginations to move from national identities to a global city perspective. "If a man be gracious and courteous to strangers," noted Francis Bacon in 1625, "he is a citizen of the world." While there is not a formal global citizenship, there is the development of a global civil discourse that contains such issues as human rights, environmental protection, social justice, peace, and shared projects of poverty reduction. There is the very real sense that we are living in one world with shared concerns and interests.

The recent economic meltdown also has raised the issue of collective economic security. It is no longer easy to mount a call for small government when small government has produced unregulated banks and unregulated oil company practices. The mortgage crises and the Gulf oil spill brought home to many the costs of a naïve commitment to the ideology of small government and big unregulated markets. Deregulation is not the simple cure for our ills. Working out a political philosophy that builds upon the security of the Keynesian/New Deal city and transcends the inequality of the neoliberal city is a major task.

There is also the collective security now demanded by the increasing threat of terrorism. Cities are massed concentrations of people and symbols. They are a target-rich environment for those who want to cause havoc and do harm. However, the provision of effective security from the possibilities of attack and terrorism can all too easily turn into forms of authoritarian control and severe restrictions on individual liberties. Collective security can all too often be a pretext for greater state power over individual lives. As Michael Bakunin noted in a speech delivered in 1867:

> There is no horror, no cruelty, sacrilege, or perjury, no imposture, no infamous transaction, no cynical robbery, no bold plunder or shabby betrayal that has not been or is not daily being perpetrated by the representatives of the states, under no other pretext than those elastic words, so convenient and yet so terrible: "for reasons of state."

We have to be wary of the threat of attack, but we also need to be vigilant about the trampling of rights and liberties. Terrorism should not be used to trump fundamental rights. Let us be acute to the danger we face but also to the values that we hold.

Community and society

We live in interesting times. Singular national and ethnic identities are being replaced by complex multicultural realities especially in the big cities of the world. This multiculturalism is an important source of cultural and economic creativity. The most successful industries draw upon global labor pools and cosmopolitanism as a source of strength and vitality for both cities and nations. However, the cultural complexity of global cities is also a source of tensions as communities with very different values and practices now share the same urban and national space. The debate in France over the wearing of the Islamic headdress is a case in point. When do national and urban mores override those of specific communities? My own view is not a knee-jerk defense of multiculturalism but a more nuanced consideration of the importance of assimilation. The European response to immigration has until very recently been concerned with cultural autonomy, leaving immigrant communities to themselves. The cultural neglect in association with a narrow definition of citizenship and restricted economic opportunities was the context for the rioting in French suburban areas and cities. I argue that a more assimilationist model is more effective for building upon democratic values. I do not believe, for example, in multiculturalism tolerant of cultures that oppress women or people with different sexual orientations. A genuine multiculturalism is only possible with a separation between church and state, an opportunity for secularism as well as religious freedoms and a basic commitment to democracy and universal human rights.

The tensions between community and society are marked in cities. People of different ethnicities and beliefs sharing the same urban space can lead to conflict. But in an overall context of tolerance for difference but commitment to basic beliefs of freedom and human rights, it can be the setting for creative dialogue. The world's cities are the setting for conflict and tension but also the stage for an achievable multicultural tolerance.

Sustainability and growth

The world's population is likely to grow and demand more goods and services. There are those who either believe there are no limits to growth or take such short-term viewpoints that long-term consequences are not considered. At the other end of the wide spectrum of opinion are the apocalyptic visions that see a world destroyed by global warming, depleted of resources, and overwhelmed by pollution overload. I am in the middle with a complex position that is concerned to extend the wealth of the world beyond a narrow range of people and societies. We need to grow somewhat in order to spread the wealth. That raises the difficult question: is sustainable growth possible? If no growth is possible then existing global inequalities are frozen in place, the poor of their world remain poor. Alternatively, we can consider smart growth with a greater commitment to a global redistribution of wealth.

An encouraging trend is the reduction of extreme poverty. From 2000 to 2015 the percentage of the worlds' population living in extreme poverty will more than half, even with the Great Recession, from 42 percent to 15 percent. The baseline is paltry; these are people living on less than $1.25 a day, many living in shantytowns with no access to clean water and safe sanitation. Yet, a mix of aid and economic growth has lifted millions from the most abject of poverty. Can the world's future economic growth be structured so that we lift more people from the depths. And can that growth be sustainable.

A smart growth that reduces, recycles, and reuses is possible. It is not easy and may take a long time. It is most attainable in the great cities of the world where people live at high densities. Cities can become and in many cases are places where more sustainable forms of living are possible. All the research tells us that the most unsustainable form of growth is low-density suburban sprawl. The carbon footprint of low-density suburbs is much higher than dense cities where public transport can replace private autos. In this sense cities are our best laboratory for finding sustainable forms of growth.

Final remarks

A number of years ago I wrote a book entitled *The Humane City*. It was subtitled *Cities as if People Matter*. It was based on my conviction that cities should be places where ordinary people can lead dignified and creative lives. I suggested that the forces against this goal were cities built as if only capital and only some people matter. To achieve more humane cities we need to ensure citizen empowerment and engagement.

A civic tradition grew up alongside urbanization. The industrial cities of the nineteenth century created pollution and conflict but were also the setting for pollution controls, public health improvements, and a host of civic associates and progressive political movements. We are in another wave of urbanization that, while containing problems, embodies similar possibilities and hopes. The cities of the world are where the sharpest tensions of collective security and individual rights, community and society, and sustainability and growth are experienced. Cities, especially the big cities of the world, are the principal setting and most important laboratory for the experience of modernity. We have many examples from around the world where the paradoxes are creatively bridged and tensions resolved. Cities are the vital dialectic that provides the opportunity to experience yet transform serious problems into creative solutions. Our urban utopia is not a destination but a journey, not an imagined endpoint but an unfolding reality, ceaselessly problematic and infinitely hopeful.

CASE STUDY 12.1 **Generational inequalities**

In an area of unprecedented and rapid change the experience and hence the attitudes and worldview of different generations, even in the same society, can differ markedly. Let me illustrate with reference to the US.

Even before the health-care reform of 2010, the US already had socialized health coverage, Medicare; it covers forty-five million people, all senior citizens. It costs around $431 billion each year, and it constitutes 57 percent of all Federal health spending. Medicare is socialized health care that no one wants to disturb. Republicans who rail against the principle of government involvement stake their criticisms well short of Medicare entitlements.

The stark difference between Medicare's coverage and how the rest of the population's health care is paid for is a clear example of generational inequity. Those born in good times get advantages over those born in bad times. Assume the (fictional) average American. Born in 1900, you experienced the Great Depression and World War II. It was only in your fifties that things began to turn around. Born in 1940 in contrast, you were carried along on the great postwar expansion of economic growth, rising incomes, and new and extended benefits. If you were white it was easy to get a job and do well. Sandwiched between tough times, those born between 1935 and 1970 are the lucky generations.

There are lucky and unlucky generations. Generational inequity really kicks in if those paying for the elderly are unlikely to see the same benefits. Older adults are advantaged because they have publicly provided pensions, health care, food stamps, housing subsidies, tax breaks, and other benefits that younger age groups do not. What is more, the younger groups, for a variety of reasons, including the relative decline in US wages and incomes due to globalization, are unlikely to receive the same level of benefits. Born in 1990, you are coming into a job market in the Great Recession, with well over a generation of stagnant incomes and increasing costs. To add insult to injury you have to work to pay for your elders to have privileged health care and generous social security payouts that you are unlikely to see for yourself. The knee-jerk defense of Medicare, while millions were uninsured and underinsured, was simply one more example of an intergenerational divide.

In 1964 Donald Horne in a bestselling book described Australia as *The Lucky Country*. In retrospect he should have entitled it more accurately as *The Lucky Cohort* as it describes a specific time when incomes were high and jobs were easily available. During this time white Australians did well. But this was not a fixed national characteristic. It was as much a function of time as space. Class, race, and gender have long been identified as sources of difference, advantage, and disadvantage. We also need to add age to that list. But therein lies an opportunity for radical shifts. The attitudes, assumptions, and practices of one generation can quickly be supplanted as younger generations with very different worldviews occupy positions of power and authority. The repeal of the don't ask, don't tell policy and the open acceptance of gays in the US military was in part a function of generational shift in attitudes. The future belongs to new generations and therein lies the possibility for a very different future, one of breaks and ruptures from this generation's status quo.

Bibliography

Ades, A. and Glaeser, E. (1995) Trade and circuses: Explaining urban giants. *Quarterly Journal of Economics*, 110, 195–227.

Alder, K. (2002) *The Measure of All Things*. New York: Simon & Schuster.

Alexander, M. L. (1967) *The Northeastern United States*. Princeton, NJ: Nostrand.

Altman, J. C. (2005) *Brokering Aboriginal Art: A critical perspective on marketing, institutions, and the state*. Geelong, Victoria: Deakin University.

Altman, J. C., Biddle, N., and Hunter, B. (2004) *Indigenous Socioeconomic Change 1971–2001: A historical perspective*. Center for Aboriginal Economic Policy Research, Discussion Paper 266. Canberra: Australian National University.

Anderson, B. (1991) *Imagined Communities: Reflections on the origins and spread of nationalism*. London: Verso.

Anderson, K. and Taylor, A. (2005) Exclusionary politics and the question of national belonging. *Ethnicities*, 5, 460–485.

Andranovitch, G., Burbank, M. J., and Heying, C. H. (2001) Olympic cities: Lessons learned from mega-event politics. *Journal of Urban Affairs*, 23, 113–131.

Anonymous (2003, June 8) California sprawl: Endless Los Angelization is winning approval. *San Diego Union Tribune*, p. G-2.

Appadurai, A. (2002) The right to participate in the work of the imagination. In J. Brouwer (Ed.) *Transurbanism*. Rotterdam: NAI Publishers, pp. 33–48.

Appiah, K. A. (2006) *Cosmopolitanism: Ethics in a world of strangers*. New York: W. W. Norton.

Armstrong. K. (2006) *The Great Transformation: The beginning of our religious traditions*. New York: Knopf.

Ashcroft, W. (Ed.) (2006) *The Post-Colonial Studies Reader* (Second ed.). London: Routledge.

Askew, M. (2002) *Bangkok: Place, practice and representation*. London: Routledge.

Austin-Broos, D. (2009) *Arrernte present, Arrernte past: Invasion, violence and imagination in indigenous central Australia*. Chicago, IL: University of Chicago Press.

Australian Competition and Consumer Commission (2007a) *Unconscionable Conduct in the Indigenous Art and Craft Sector*. Canberra: Commonwealth of Australia.

Australian Competition and Consumer Commission (2007b) *Your Consumer Rights: Indigenous art and craft*. Canberra: Commonwealth of Australia.

Australian Senate: Standing Committee on Environment, Communications, Information Technology and the Arts (2007) *Indigenous Art: Securing the future, Australia's indigenous visual arts and crafts sector*. Canberra: Senate Printing Unit.

Bardon, G. and Bardon, J. (2004) *Papunya: A place made after the story*. Carlton, MN: Miegunyah Press.

Barney, R. K. (2002) *Selling the Five Rings: The International Olympic Committee and the rise of olympic commercialism*. Salt Lake City, UT: University of Utah Press.

Bassett, K. A. (2004) Walking as an aesthetic practice and a critical tool. *Journal of Geography in Higher Education*, 28, 397–410.

Baudelaire, C. (1965) *Art in Paris, 1845–1862*, (Jonathan Mayne, Trans.). Oxford, UK: Phaidon.

Baudelaire, C. (1970) *The Painter of Modern Life: And other essays* (Jonathan Mayne, Trans.). Oxford, UK: Phaidon.

Bauman, Z. (1990) Modernity and ambivalence. *Theory, Culture & Society*, 7(2), 143–169.

Bauman, Z. (1992) *Intimations of Postmodernity*. London: Routledge.

Bauman, Z. (2005) *Liquid Life*. Cambridge, UK: Polity Press.

Baumgartner, K. (2008) Constructing Paris: *Flânerie*, female spectatorship, and the discourses of fashion in Französische Miscellen (1803). *Monatshefte, 100*, 351–368.

Beaverstock, J. V. (2002) Transnational elites in global cities: British expatriates in Singapore's financial district. *Geoforum*, 33, 525–538.

Beaverstock, J. V. (2007) World city networks "from below": International mobility and inter-city relations in the global investment banking industry. In P. J. Taylor, B. Derudder, P. Saey, and F. Witlox (Eds.) *Cities in Globalization*. New York: Routledge, pp. 50–68.

Beck, U., Giddens, A., and Lash, S. (1994) *Reflexive Modernization: Politics, tradition and aesthetics in the modern social order*. Cambridge, UK: Polity Press.

Beck, U. and Sznaider, N. (2006) Unpacking cosmopolitanism for the social sciences: A research agenda. *British Journal of Sociology*, 57, 1–23.

Benjamin, W. (1999) *The Arcades Project*. Cambridge, MA: Harvard University Press.

Benjamin, W. and Jennings, M. W. (2006) *The Writer of Modern Life: Essays on Charles Baudelaire*. Cambridge, MA: Belknap Press.

Benton-Short, L., Price, M. D., and Friedman S. (2005) Globalization from below: The ranking of global immigrant cities. *International Journal of Urban and Regional Research*, 29, 945–959.

Benton-Short, L. and Short, J. R. (2008) *Cities and Nature*. London: Routledge.

Berg, L., Pol, P., Mingardo, G., and Speller, C. (Eds.) (2006) *The Safe City*. Aldershot & Burlington: Ashgate.

Bergdoll, B. and Dickerman, L. (eds) (2009) *Bauhaus: Workshops for modernity*. New York: Museum of Modern Art.

Bergen, K. (2007, April 1) Cost, profits rarely clear-cut for host city. *Chicago Tribune*, p. 1.

Beriatos, E. and Gospodini, A. (2004) "Glocalizing" urban landscapes: Athens and the 2004 Olympics. *Cities*, 21, 187–204.

Berke, E., Koepsell, T., Moudon, V., Hoskins, R., and Larson, E. (2007) Association of the built environment with physical activity and obesity in older persons. *American Journal of Public Health*, 97, 486–492.

Berman, M. (1982) *All that Is Solid Melts into Air: The Experience of Modernity*. New York: Simon & Schuster.

Berman, M. (1992) Why modernism still matters. In S. Lash and J. Friedmann (eds), *Modernity and Identity*. Oxford, UK: Blackwell, pp. 33–58.

Berube, A. and Tiffany, T. (2005) The shape of the curve: Household income distributions in U.S. cities, 1979–99. In A. Berube, B. Katz, and R. E. Lang (Eds.) *Redefining Urban and Suburban America: Evidence from census 2000, Volume 2*. Washington, DC: Brookings Institution Press, pp. 195–243.

Bessire, M. (2009) *Stairway to Heaven: From Chinese streets to monuments and skyscrapers*. Hanover and London: University Press of New England.

Bishai, D., Hyder, A., Ghaffar, A., Morrow, R., and Kobusingye, O. (2003) Rates of public investment for road safety in developing countries: Case studies of Uganda and Pakistan. *Health Policy and Planning*, 18, 232–235.

Blaise, C. (2000) *Time Lord: Sir Sandford Fleming and the creation of standard time*. London: Weidenfeld & Nicholson.

Blanchard, M. (1985) *In Search of the City: Engels, Baudelaire, Rimbaud*. Saratoga, CA: Anma Libri.

Blanchard, M. (1993) Between autobiography and ethnography: The journalist as anthropologist. *Diacritics*, 23(4), 72–81.

Bogue, D. (1951) *State Economic Areas*. Washington, DC: US Government Printing Office.

Borchert, J. R. (1992) *Megalopolis: Washington D.C. to Boston*. New Brunswick, NJ: Rutgers University Press.

Boyle, R. (1949, October 17) Report on communist China. *Life*, pp. 129–142.

Braudel, F. (1981) *The Structures of Everyday Life*. New York: Harper & Row.

Brinkhoff, T. (2010) The principal agglomerations of the world. Retrieved July 23, 2010 from www.citypopulation.de.

Brooks, D. (2003) *A Town Like Mparntwe*. Alice Springs: Jukurrpa Books.

Broudehoux, A-M. (2007) Spectacular Beijing: The conspicuous consumption of an Olympics metropolis. *Journal of Urban Affairs*, 29, 383–400.

Brownson, R., Baker, E., Housemann, R., Brennan, L., and Bacak, S. (2001) Environmental and policy determinants of physical activity in the United States. *American Journal of Public Health*, 91, 1995–2003.

Brunet, F. (2005) *The Economic Impact of the Barcelona Olympic Games, 1986–2004*. Barcelona: Centre d'Estudis Olimpics UAB. Retrieved March 5, 2009 from http://olympicstudies.uab.es/pdf/wp084_eng.

Buchan, J. (2003) *Capital of the Mind: How Edinburgh changed the world*. John Murray: London.

Burbank, M. (2001) *Olympic Dreams: The impact of mega-events on local politics*. Boulder, CO: Lynne Rienner.

Burbank, M. J., Heying, C. H., and Andranovich, G. (2000) Antigrowth politics or piecemeal resistance: Citizen opposition to Olympic-related economic growth. *Urban Affairs Review*, 35, 334–357.

Buruma, I. and Margalit, A. (2004) *Occidentalism: The West in the eyes of its enemies*. New York: Penguin.

Butler, T. (2006) A walk of art: The potential of the sound walk as cultural practice in cultural geography. *Social and Cultural Geography*, 7, 889–908.

Campanella, T. (2008) *The Concrete Dragon: China's urban revolution and what it means for the world*. New York: Princeton Architectural Press.

Campbell, L. (2006) *Darby: One hundred years of life in a changing culture*. Sydney: ABC Books.

Canclini, N. and Chiappari, C. L. (1995) *Hybrid Cultures: Strategies for entering and leaving modernity*. Minneapolis, MN: University of Minnesota Press.

Castells, M. (1996) *The Rise of Network Society*. Oxford, UK: Blackwell.

Celis, A., Gómez, Z., Martínez-Sotomayor, A., Arcila, L., and Villaseñor, M. (2003) Family characteristics and pedestrian injury risk in Mexican children. *Injury Prevention*, 9, 58–61.

Center on Housing Rights and Evictions (2007) *Fair Play for Housing Rights: Mega-events, olympic games and housing rights*. Retrieved March 5, 2009 from www.cohre.org/mega-events.

Cervero, R. and Duncan, M. (2003) Walking, bicycling, and urban landscapes: Evidence from the San Francisco Bay Area. *American Journal of Public Health*, 93, 1478–1483.

Cha, A. E. (2009, March 4) In China, despair mounting among migrant workers. *The Washington Post*, p. A10.

Chandler, J. and Gilmartin, K. (2005) *Romantic Metropolis: The urban scene of British culture, 1780–1840*. Cambridge, MA: Cambridge University Press.

Chang, T. C., Huang, S., and Savage, V. R. (2004) On the waterfront: globalization and urbanization in Singapore. *Urban Geography*, 25, 413–436.

Chen, X. (Ed.) (2009) *Shanghai Rising*. Minneapolis, MN: University of Minnesota Press.

Christie, I. (2006) Mass-market modernism. In Wik, C. (Ed.), *Modernism: Designing a new world*. London: V & A Publications, pp. 375–414.

Cockburn, J. (2005) *Simmel, Ninotchka and the Revolving Door*. Paper presented at the Proceedings of Cosmopolitanism and Place: The Design Resistance Conference, Sydney, Australia.

Crandall, J., Bhalla, K., and Madeley, N. (2002) Designing road vehicles for pedestrian protection. *British Medical Journal*, 324, 1145–1148.

D'Souza, A. and McDonough, T. (2006) *The Invisible Flâneuse?: Gender, public space, and visual culture in nineteenth-century Paris*. New York: Manchester University Press.

Dahl, R. (2004) Vehicular manslaughter: The global epidemic of traffic death. *Environmental Health Perspectives*, 112, A628–A631.

David, B. (2008) Archeology and the dreaming: Toward an archeology of ontology. In I. Lilly (Ed.) *Archaeology of Oceania: Australia and the Pacific Islands*. Oxford, UK: Blackwell, pp. 48–68.

Davidson, R. (1980) *Tracks*. London: Jonathan Cape.

Debord, G. (1973) *Society of the Spectacle*. Detroit, MI: Black & Red.

De Certeau, M. (1984) *The Practice of Everyday Life*. Berkeley, CA: University of California Press.

Deleuze, G. (1995) *Negotiations*. New York: Columbia University Press.

Deleuze, G. and Guatarri, F. (1987) *A Thousand Plateaus: Capitalism and schizophrenia*. Minneapolis, University of Minnesota Press.

Dicken, P. (2003) *Global Shift: Reshaping the global economic map in the 21st century*. London: Sage.

Dickens, C. and Horne, R. (1851, July 5) The Great Exhibition and the little one. *Household Words*, pp. 356–360.

DiMaggio, C., Durkin, M., and Richardson, L. (2006) The association of light trucks and vans with paediatric pedestrian deaths. *International Journal of Injury Control and Safety Promotion*, 13, 95–99.

Dong, S. (2000) *Shanghai: The Rise and Fall of a Decadent City.* New York: HarperCollins.

Dunbar-Hall, P. and Gibson, C. (2005) *Deadly Sounds, Deadly Places: Contemporary music in Australia.* Seattle, WA: University of Washington Press.

Durkin, M., Laroque, D., Lubman, I., and Barlow, B. (1999) Epidemiology and prevention of traffic injuries to urban children and adolescents. *Pediatrics*, 3, 1–8.

Elder, B. (2003) *Blood on the Wattle* (Third ed.). Sydney: New Holland.

Elsheshtawy, Y. (2008) Transitory sites: Mapping Dubai's "forgotten" urban spaces. *International Journal of Urban and Regional Research*, 32, 968–988.

Engwicht, D. (1999) *Street Reclaimimg: Creating livable streets and vibrant communities.* Annandale, VA: Pluto Press.

Essex, S. J. and Chalkley, B. S. (1998) The Olympics as a catalyst of urban change. *Leisure Studies*, 17, 187–206.

Evans, L. (2003) A new traffic safety vision for the United States. *American Journal of Public Health*, 93, 1384–1386.

Fahn, J. (2003) *A Land on Fire: The environmental consequences of the Southeast Asia boom.* Boulder, CO: Westview.

Fan, C. (2008a, July 12) As Beijing Olympics near, homes and hope crumble. *The Washington Post*, p. A6.

Fan, M. (2008b, January 26) Shanghai's middle class launches quiet, meticulous revolt. *The Washington Post*, pp. A1, A14.

Farley, T., Meriwether, R., Baker, E., Watkins, L., Johnson, C., and Webber, L. (2007) Safe play spaces to promote physical activity in inner city children: Results from a pilot study of an environmental intervention. *American Journal of Public Health*, 97, 1625–1631.

Featherstone, M. (1998) The *flâneur*, the city and virtual public life. *Urban Studies*, 35(5/6), 909–925.

Featherstone, M., Lash, S., and Robertson, R. (1995) *Global Modernities*. London: Sage.

Featherstone, M., Thrift, N., and Urry, J. (Eds.) (2005) *Automobilities*. London: Sage.

Ferguson, P. P. (1994) The *flâneur* on and off the streets of Paris. In K. Tester (Ed.), *The Flâneur.* London: Routledge, pp. 23–42.

Findling, J. and Pelle, K. D. (2008) *Encyclopedia of World's Fairs and Expositions.* Jefferson, NC: McFarland.

Finnegan, W. (2000) After Seattle. *The New Yorker* April 17, pp. 40–51.

Flint, R. (1972) *Marinetti: Selected writings*. New York: Farrar, Straus & Giroux.

Florida, R. (2002) *The Rise of the Creative Class.* New York: Basic Books.

Foster, D., Mitchell, J., Ulrik, J., and Williams, R. (2005) *Population and Mobility in Town Camps of Alice Springs.* Alice Springs: Tangentyere Council Research Unit, Desert Knowledge Cooperative Research Center.

Foucault, M. (1967) *Madness and Civilization: A history of insanity in the age of reason.* London: Tavistock.

Freeman, V., Shanahan, E., and Guild, P. (2004) *Reducing Mortality from Motor-Vehicle Crashes for Children 0 Through 14 Years of Age: Success in New York and North Dakota*. Cecil G. Sheps Center for Health Services Research. University of North Carolina at Chapel Hill.

Friedman, S. (2005) Planetary modernism and the modernities of empires and new nations. Oral presentation. Madison, WI: University of Wisconsin, Humanities Center.

Friedman, T. (2005) *The World is Flat*. New York: Farrar, Straus & Giroux.

Frisby, D. (1981) *Sociological Impressionism: A reassessment of Georg Simmel's social theory*. London: Heinemann.

Frisby, D. (1988) *Fragments of Modernity: Theories of modernity in the work of Simmel, Kracauer, and Benjamin*. Cambridge, MA: MIT Press.

Gamble, J. (2003) *Shanghai in Transition*. London: Routledge.

Gandelsonas, M. (Ed.) (2002) *Shanghai Reflections: Architecture, urbanism and the search for an alternative modernity*. New York: Princeton Architectural Press.

Gandy, M. (2008) Landscapes of disaster: water. Modernity and urban fragmentation in Mumbai. *Environment and Planning A*, 40, 108–130.

Garreau, J. (1991) *Edge Cities: Life on the new frontier*. New York: Doubleday.

Gaulton, R. (1981) Political mobilization in Shanghai in 1949–1951. In C. Howe (Ed.) *Shanghai: Revolution and development in an Asian metropolis*. Cambridge: Cambridge University Press, pp. 35–65.

Gay, P. (1969) *The Enlightenment: The science of freedom*. New York: A. A. Knopf.

Gelder, K. and Jacobs, J. M. (1998) *Uncanny Australia: Sacredness and identity in a postcolonial nation*. Carlton: Melbourne University Press.

Genocchio, B. (2008) *Dollar Dreaming: Inside the Aboriginal art world*. Prahan: Hardie Grant.

Gerlach, N. and Hamilton, S. N. (2004) Preserving self in the city of the imagination. *Canadian Review of American Studies*, 34, 115–134.

Gessen, K. (2010, August 2) Stuck: The meaning of the city's traffic nightmare. *The New Yorker*, 24–28.

Giddens, A. (1991) *Modernity and Self-identity: Self and society in the late modern age*. Cambridge, MA: Polity Press.

Gill, S. D. (1998) *Storytracking: Texts, stories and histories in Central Australia*. Oxford, UK: Oxford University Press.

Gilloch, G. (2002) *Walter Benjamin, Critical Constellations*. Cambridge, MA: Polity Press.

Gilloch, G. (2007) Urban optics: Phantasmagoria and the city in Benjamin and Kracauer. *New Formations*, 61, 115–131.

Goebel, R. J. (2001) *Benjamin Heute: Grossstadtdiskurs, postkolonialität und Flânerie zwischen den kulturen*. Munich: Iudicium.

Gold, J. and Gold, M. (2005) *Cities of Culture: Staging international festivals and the urban agenda*. Aldershot: Ashgate.

Gold, J. and Gold, M. (Eds.) (2007) *Olympic Cities City Agendas and the World's Games, 1896–2012*. London: Routledge.

Goldberger, P. (2006, December–January) Shanghai surprise. *The New Yorker*, pp. 144–145.

Golubchikov, O. (2010) World-city-entrepreneurialism: Globalist imaginaries, neoliberal geographies and the production of new St Petersburg. *Environment and Planning A*, 42, 626–643.

Goodman, B. (1995) *Native Place, City and Nation: Regional networks and identities in Shanghai, 1853–1937.* Berkeley, CA: University of California Press.

Goot, M. and Rowse, T. (2007) *Divided Nation: Indigenous affairs and the imagined public.* Carlton: Melbourne University Publishing.

Gottmann, J. (1957) Megalopolis: Or the urbanization of the northeastern seaboard. *Economic Geography*, 33, 189–200.

Gottmann, J. (1961) *Megalopolis.* New York: Twentieth Century Fund.

Grant, R. (2009) *Globalizing City: The urban and economic transformation of Accra, Ghana.* Syracuse, NY: Syracuse University Press.

Grossman, D. (2000) The history of injury control and the epidemiology of child and adolescent injuries. Retrieved June 2, 2008 from www.futureofchildren.org.

Gumbrecht, H. and Marrinan, M. (2003) *Mapping Benjamin: The work of art in the digital age.* Palo Alto, CA: Stanford University Press.

Habermas, J. (1992) *Postmetaphysical Thinking.* Cambridge, UK: Polity.

Hägerstrand, T. (1970) What about people in regional science? *Papers of the Regional Science Association*, 24, 7–21.

Hall, P. G. (1973) *The Containment of Urban England.* London: Allen & Unwin.

Hall, P. G. (1998) *Cities in Civilization.* London: Weidenfeld & Nicholson.

Hall, T. and Hubbard, P. (Eds.) (1998) *The Entrepreneurial City: Geographies of politics, regime and representation.* Chichester: John Wiley.

Hamilton, A. (1990) Fear and desire: Aborigines, Asians and the national imaginary. *Australian Cultural History*, 9, 14–35.

Han, S. S. (2005) Global city making in Singapore: a real estate perspective. *Progress in Planning* 64, 69–175.

Hanlon, B. (2009) *Once the American Dream: Inner-ring suburbs of the metropolitan United States.* Philadelphia, PA: Temple University Press.

Hardt, M. and Negri, A. (2000) *Empire.* Cambridge, MA: Harvard University Press.

Harris, A. (2008) From London to Mumbai and back again: Gentrification and public policy in comparative perspective. *Urban Studies*, 45, 2407–2428.

Harris, C. (1982) The urban and industrial transformation of Japan. *Geographical Review*, 72, 50–89.

Harvey, D. (1989a) From managerialism to entrepreneurialism: The transformation in urban governance in late capitalism. *Geografiska Annaler Series B*, 71, 3–17.

Harvey, D. (1989b) *The Condition of Postmodernity.* Oxford, UK: Basil Blackwell.

Harvey, P. (1996) *Hybrids of Modernity: Anthropology, the nation state and the universal exhibition.* London. Routledge.

Hass-Klau, C. (1990) *The Pedestrian and City Traffic.* London: Belhaven.

Hayes v. Northern Territory (1999) Federal Court of Australia 1248. Retrieved February 5, 2010 from www.austlii.edu.au/au/cases/cth/federal_ct/1999/1248.html.

He, S. and Wu, F. (2007) Neighborhood changes and residential differentiation in Shanghai. In F. Wu (Ed.), *China's Emerging Cities.* New York: Routledge, pp. 295–209.

Herman, A. (2001) *How the Scots Invented the Modern World.* Random House: New York.

Hewson, P. (2004) Deprived children or deprived neighborhoods? A public health approach to the investigation of links between deprivation and injury risks or specific reference to child road safety in Devon county, UK. *BioMed Central Public Health*, 4, 4–15.

Highmore, B. (2004) Homework. *Cultural Studies*, 18, 306–327.

Hillman, M. (1993) *Children, Transport and the Quality of Life*. London: Policy Studies Institute.

Himmelfarb, G. (2004) *The Roads to Modernity*. New York: Knopf.

Hinder, E. M. (1942) *Life and Labor in Shanghai's International Settlement.* Shanghai: Shanghai Municipal Council.

Honig, E. (1992) *Sisters and Strangers: Women in the Shanghai cotton mills, 1919–1949.* Stanford, CA: Stanford University Press.

Horne, J. and Manzenreiter, W. (2006) An introduction to the sociology of sports mega-events. *Sociological Review*, 54, 1–24.

Hotchkiss, J. E., Moore, R. R. and Zobay, S. M. (2003) Impact of the 1996 summer Olympic Games on employment and wages in Georgia. *Southern Economic Journal*, 69, 691–704.

Huang, T. (2004) *Walking Between Slums and Skyscrapers: Illusions of open space in Hong Kong, Tokyo, and Shanghai.* Hong Kong: Hong Kong University Press.

Hsing, Y. (2009) *The Great Urban Transformation: Politics of land and property in China.* New York: Oxford University Press.

Ibrahim, Y. (2007) The technological gaze: Event construction and the mobile body. *M/C Journal*, 10(1). Retrieved October 1, 2010 from http://journal.media-culture.org.au/0703/03-ibrahim.php.

IOC (International Olympic Committee) (2007) The Top Program. Retrieved from www.olympic.org/uk/organisation/facts/program/sponsors_uk.asp.

Ishaque, M. and Noland, R. (2006) Making roads safe for pedestrians or keeping them out of the way? *The Journal of Transport History*, 27, 115–137.

Israel, J. (2001) *Radical Enlightenment: Philosophy and the making of modernity 1650–1750.* Oxford, UK: Oxford University Press.

Jacobs, G., Aeron-Thomas, A. and Astrop, A. (Eds.) (2000) *Estimating Global Road Fatalities.* London: Transport Research Laboratory.

Jacobs, J. (1993) The city unbound: Qualitative approaches to the city. *Urban Studies*, 30(4), 827–848.

Jefferson, M. (1939) The law of the primate city. *Geographical Review*, 29, 226–232.

Jennings, A. (1996) *The New Lords of the Rings*. New York: Simon & Schuster.

Jennings, A. and Sambrook, C. (2000) *The Great Olympic Swindle*. New York: Simon & Schuster.

Jensen, O. (2006) Facework, flow and the city: Simmel, Goffman, and mobility in the contemporary city. *Mobilities*, 1, 143–165.

Johnson, S. (2006) *The Ghost Map: The story of London's most terrifying epidemic and how it changed science, cities and the modern world*. New York: Penguin.

Johnston, T. (2008, February 13) Australia says "sorry" to Aborigines for mistreatment. *New York Times*, p. 12.

Jonas, A. and Wilson, D. (Eds.) (1999) *The Urban Growth Machine: Critical perspectives two decades later*. Albany, NY: SUNY Press.

Kahn, J. S. (2003) Anthropology as cosmopolitan practice? *Anthropological Theory*, 3, 403–415.

Kalliney, P. J. (2002) Globalization, postcoloniality, and the problem of literary studies in The Satanic Verses. *Modern Fiction Studies*, 48, 50–82.

Kasimati, E. (2003) Economic aspects and the Summer Olympics: A review of related research. *International Journal of Tourism Research*, 5, 433–444.

Kawachi, I. and Berkman, L. (Eds.) (2003) *Neighborhoods and Health.* New York: Oxford University Press.

Keen, I. (2004) *Aboriginal Economy and Society: Australia at the threshold of colonization.* London: Oxford University Press.

Keohane, K. (2002) The revitalization of the city and the demise of Joyce's utopian modern subject. *Theory, Culture & Society*, 19, 29–49.

Kern, S. (1983) *The Culture of Time and Space 1880–1918.* Cambridge, MA: Harvard University Press.

Kim, K-J. (Ed.) (2003) *Seoul, Twentieth Century: Growth and change of the last 100 years.* Seoul: Seoul Development Institute.

Kim, Y-H. and Short, J. R. (2008) *Cities and Economies.* New York: Routledge.

Kimber, R. G. (1991) *The End of the Bad Old Days: European settlement in Central Australia 1871–1894.* Darwin: State Library of the Northern Territory.

King, A. (1997) *Culture, Globalization, and the World-System: Contemporary conditions for the representation of identity.* Minneapolis, MN: University of Minnesota Press.

King, A. D. (2004) *Spaces of Global Cultures: Architecture, urbanism, identity.* London: Routledge.

King, A. D. (2009) Unravelling modernities. In *Humanistic Reflections for the Humane City.* Incheon: Incheon Conference for Urban Humanities, pp. 25–32.

Knight, M. and Chan, D. (2010) *Shanghai: Art of the City.* San Francisco, CA: Asian Art Museum.

Kong, L. and Yeoh, B. S. A. (2003) *The Politics of Landscape in Singapore.* Syracuse, NY: Syracuse University Press.

Kremmer, J. (2006, June 7) Australia aims to protect a $149 million art industry. *The Christian Science Monitor*, p. 4.

Krisberg, K. (2006) Built environment adding to burden of childhood obesity. *Nation's Heath*, 36, 1–27.

Krishnamurthy, A. (2000) Friedrich Engels in industrial Manchester. *Victorian Literature and Culture*, 28, 427–448

Krugman, P. (1996) Urban concentration: The role of increasing returns and transportation costs. *International Regional Science Review*, 19, 5–30.

Kuan, S. and Rowe, P. G. (Eds.) (2004) *Shanghai: Architecture and urbanism for modern China.* Munich: Prestel.

Lake, R. (2003) Gottmann forty years on. Unpublished paper based on panel discussion at Association of American Geographers Mid States Geography Conference.

Lambiri, D. (2005, May 30) *The Olympic Village of Barcelona: Urban residential development and socio-economic impact.* Paper presented to 5th International Conference on Sport and Culture. Athens, Greece.

Lang, R. E. (2003) *Edgeless Cities: Exploring the elusive metropolis.* Washington, DC: Brookings Institution Press.

Lang, R. E. and Dhavale, D. (2005) Beyond Megalopolis: Exploring America's new "megapolitan" geography. *Metropolitan Institute of Virginia Tech. Census Report Series*, 5, 1–35.

Lash, S. (1999) *Another Modernity, a Different Rationality.* London: Wiley-Blackwell.

Latham, P. and McCormack, D. (2008) Speed and slowness. In T. Hall, P. Hubbard, and J. Short (Eds.), *The Sage Companion to the City.* London: Sage, pp. 301–317.

Latour, B. (1993) *We Have Never Been Modern*. Cambridge, MA: Harvard University Press.

Lauster, M. (2007) Walter Benjamin's myth of the *flâneur*. *Modern Language Review*, 102, 139–156.

Lee, L. O. (1999) *Shanghai Modern: The flowering of a new urban culture in China, 1930–1945*. Cambridge, MA: Harvard University Press.

Lenskyj, H. J. (2000) *Inside the Olympic Industry*. Albany, NY: SUNY Press.

Lenskyj, H. J. (2002) *The Best Olympics Ever? Social impacts of Sydney 2000*. New York: State University of New York Press.

Levinson, M. (2006) *The Box: How the shipping container made the world economy bigger*. Princeton, NJ: Princeton University Press.

Linsky, A. (1965) Some generalizations concerning primate cities. *Annals of Association of American Geographers*, 55, 506–513.

Llosa, A. (2005) *Liberty for Latin America*. New York: Farrar, Straus & Giroux.

Logan, J. and Molotch, H. (1987) *Urban Fortunes: The political economy of place*. Berkeley, CA: University of California Press.

London, B. (1979) Internal colonialism in Thailand. *Urban Affairs Review*, 4, 485–513.

Lopez, R. (2004) Urban sprawl and risks of being overweight or obese. *American Journal of Public Health*, 94, 1574–1579.

Luke, R. (2005) The Phoneur: Mobile commerce and the digital pedagogies of the wireless web. In P. P. Trifonas (Ed.), *Communities of Difference: Culture, language, technology*. New York: Palgrave Macmillan, pp. 185–204.

Lynch, K. (1972) *What Time Is this Place?* Cambridge, MA: MIT Press.

Mabo and Others v. Queensland (No. 2) (1992) High Court of Australia 23. Retrieved February 6, 2010 from www.austlii.edu.au/au/cases/cth/HCA/1992/23.html.

MacGillis, A. (2006, April 25) Ties to far-flung homes drive commuters to great lengths. *The Washington Post*, pp. A1, A12.

Markus, A. (1990) *Governing Savages*. North Sydney: Allen & Unwin.

Martin, T. (1880) *The Life of His Royal Highness the Prince Consort*, Volume 2. New York: D. Appleton & Co.

Marx, P. (2008, July 21) Buy Shanghai. *The New Yorker*, pp. 40–44.

Massey, D. (2007) *World City*. Cambridge, UK: Polity Press.

Matheson, V. A. (2002) Upon further review: Sports events economic impact studies. *The Sports Journal*, 5(2). Retrieved March 5, 2009 from www.thesportjournal. org/2002Journal/Vol5-No1/studies.htm.

McGee, T. G., Lin, G. C. S., Marton, A. M., Wang, M. Y. L., and Wu, J. (2007) *China's Urban Space: Development under market socialism*. London: Routledge.

McKay, M. and Plumb, C. (2001) *Reaching Beyond the Gold*. London: LaSalle Investment Management.

Meagher, S. (2007) Philosophy in the streets: Walking the city with Engels and de Certeau. *City*, 11, 7–20.

Mehta, S. (1964) Some demographic and economic correlates of primate cities: A case for revaluation. *Demography*, 1, 136–147.

Mendelsohn, J.A. (2003) The microscopist of modern life. *Osiris*, 18, 150–170.

Messinger, G. S. (1985) *Manchester in the Victorian Age: The half-known city*. Manchester: Manchester University Press.

Michaels, E. (1994) *Bad Aboriginal Art: Tradition, media and technological horizons*. Minneapolis, MN: University of Minnesota Press.

Miele, M. (2008) CittaSlow: Producing slowness against the fast life. *Space and Policy*, 12, 135–156.

Mitchell, T. (1988) *Colonizing Egypt.* Cambridge, UK: Cambridge University Press.

Mohan, D. (2002) Road safety in less-motorized environments: Future considerations. *International Journal of Epidemiology* 31, 527–532.

Mohan, D., Tiwari, G., Khayesi, M., and Muyia-Nafukho, F. (2006) *Road Traffic Injury Prevention Training Manual.* Geneva: World Health Organization.

Moomaw, R. and Alwosabi, M. (2004) An empirical analysis of competing explanations of urban primacy evidence from Asia and the Americas. *Annals of Regional Science*, 38, 149–171.

Morphy, H. (2008) *Becoming Art.* Sydney: UNSW Press.

Mutebi, A. M. (2004) Recentralising while decentralising: Center–local relations and "CEO" governors in Thailand. *The Asia Pacific Journal of Public Administration*, 26, 33–53.

Myers, F. R. (2002) *Painting Culture: The making of an Aboriginal high art.* Durham, NC: Duke University Press.

Newman, K. and Wyly, E. K. (2006) The right to stay put revisited: Gentrification and resistance to displacement in New York City. *Urban Studies*, 43, 23–57.

Newman, P. (2007) "Back the bid": The 2012 Summer Olympics and the governance of London. *Journal of Urban Affairs*, 29, 255–268.

Normark, D. (2007) *Enacting Mobility: Studies in the nature of road-related social interaction.* Göteburg: Göteburg University.

Norton, P. D. (2008) *Fighting Traffic: The dawn of the motor age in the American city.* Cambridge, MA: MIT Press.

Nowicka, M. (2006) Mobility, space and social structuration in the second modernity and beyond. *Mobilities*, 1(3), 411–435.

Nylund, K. (2001) Cultural analyses in urban theory of the 1990s. *Acta Sociologica*, 44(3), 219–230.

O'Flaherty, B. (2005) *City Economics.* Cambridge, MA: Harvard University Press.

Old, K. and Yeung, H. W-C. (2004) Pathways to global city formation: a view from the developmental city-state of Singapore. *Review of International Political Economy*, 11, 489–521.

Ooi, C-S. (2008) Reimagining Singapore as a creative nation: The politics of place branding. *Place Branding and Public Diplomacy*, 4, 287–302.

Osborn, F. J. and Whittick, A. (1963) *The New Towns: The answer to Megalopolis.* London: McGraw-Hill.

Ouroussof, N. (2009, November 6) Finding a bit of animal house in the Bauhaus. *New York Times*, p. C25.

Packer, G. (2006, November 13) The megacity: Decoding the chaos of Lagos. *The New Yorker*, pp. 62–75.

Parker, C. H. (2010) *Global Interactions in the Early Modern Age (1400–1800).* Cambridge, UK: Cambridge University Press.

Parker, I. (2008, November 10) The bright side. *The New Yorker*, pp. 52–63.

Parsons, D. L. (2000) *Streetwalking the Metropolis: Women, the city and modernity.* Oxford, UK: Oxford University Press.

Pearlman, J. (2000) Joseph Hudnut and the unlikely beginnings of post-modernism urbanism at the Harvard Bauhaus. *Planning Perspectives*, 15, 201–239.

Peck, J. (2005) Struggling with the creative class. *International Journal of Urban and Regional Research*, 29, 740–770.

Peden, M., Scurfield, R., Sleet, D., Mohan, D., Hyder, A., Jarawan, E., and Mathers (2004) *World Report on Road Traffic Injury Prevention: Summary*. Geneva: World Health Organization.

Pell, C. (1966) *Megalopolis Unbound: The supercity and the transportation of tomorrow*. New York: Praeger.

Perkins, H. and Fink, H. (Eds.) (2000) *Papunya Tula: Genesis and genius*. Sydney: Art Gallery of New South Wales.

Perrow, C. (1999) *Normal Accidents: Living with high risk technologies*. Princeton, NJ: Princeton University Press.

Philo, C. (1999) Edinburgh, Enlightenment, and the geographies of unreason. In D. N. Livingstone and C.W. J. Withers (Eds.) *Geography and Enlightenment*. Chicago, IL: University of Chicago Press, pp. 372–398.

Pile, S. (1996) *The Body and the City: Psychoanalysis, space, and subjectivity*. London: Psychology Press.

Pilkington, P. (2000) Reducing the speed limit to 20 mph in urban areas. *British Medical Journal*, 320, 1160.

Pinder, D. (2002) In defense of utopian urbanism: Imagining cities after the "end of utopia." *Geografiska Annaler B*, 84, 229–241.

Pink, B. and Allbon, P. (2008) *The Health and Welfare of Australia's Aboriginal and Torres Strait Islander Peoples*. Canberra: ABS.

Pink, S., Hubbard, P., O'Neill, M., and Radley, A. (2010) Walking across the disciplines: From ethnography to arts practice. *Visual Studies*, 25, 1–7.

Popke, J. (2009) Geography and ethics: Non-representational encounters, collective responsibility and economic difference. *Progress in Human Geography*, 33, 81–90.

Povinelli, E. A. (2002) *The Cunning of Recognition: Indigenous alterities and the making of Australian multiculturalism*. Durham, NC: Duke University Press.

Pow, C-P. (2009) *Gated Communities in China*. London: Routledge.

Pred, A. (1995) *Recognizing European Modernities: A montage of the present*. London. Routledge.

Preuss, H. (2004) *The Economics of Staging the Olympics*. Cheltenham: Edward Elgar.

Pridmore, J. (2008) *Shanghai: The architecture of China's great urban center*. New York: Abrams.

Pucher, J. and Dijkstra, L. (2003) Promoting safe walking and cycling to improve public health: Lessons from the Netherlands and Germany. *American Journal of Public Health*, 93, 1509–1516.

Puentes, R. and Warren, D. (2006) *One-Fifth of America: A comprehensive guide to America's first suburbs*. Washington, DC: Brookings Institution Press.

Putnam, S. H. (1975) *An Empirical Model of Regional Growth: With an application to the Northeast Megalopolis*. Philadelphia, PA: Regional Science Institute.

Rao, V. (2007) Proximate distances: The phenomenology of density in Mumbai. *Built Environment*, 33, 227–248.

Ren, X. (2008) Architecture and nation building in an age of globalization: Construction of the National Stadium of Beijing for the 2008 Olympics. *Journal of Urban Affairs*, 30, 175–190.

Renson, R. and Hollander, M. den (1997) Sport and business in the city: The Antwerp Olympic Games of 1920 and the urban elite. *Journal of Olympic Studies, VI*, 73–84.

Resina, J. and Ingenschay, D. (2003) *After-Images of the City*. Ithaca, NY: Cornell University Press.

Retting, R., Ferguson, S., and McCartt, A. (2003) A review of evidence-based traffic engineering measures designed to reduce pedestrian–motor vehicle crashes. *American Journal of Public Health*, 93, 1456–1463.

Reynolds, H. (1981) *The Other Side of the Frontier: An interpretation of the Aboriginal response to the invasion and settlement of Australia.* Townsville, QLD: History Department, James Cook University.

Reynolds, H. (2001) *An Indelible Stain? The question of genocide in Australia's history.* Ringwood, VIC: Viking.

Richburg, K. B. (2010, August 15) In China, chafing under ancient permits. *The Washington Post*, p. A8.

Richter, E. D., Friedman, L. S., Berman, T., and Avraham, R. (2005) Death and injury from motor vehicle crashes: A tale of two countries. *American Journal of Preventive Medicine*, 29, 440–449.

Richter, G. (2002) *Benjamin's Ghosts: Interventions in contemporary literary and cultural theory*. Palo Alto, CA: Stanford University Press.

Rigg, J. (1995) *Counting the Costs: Economic growth and environmental change in Thailand*. Singapore: Institute of Southeast Asian Studies.

Roche, M. (2006) Mega-events and modernity revisited: Globalization and the case of the Olympics. *Sociological Review*, 54, 25–40.

Rotbard, S. (2005) *White City, Black City*. Tel Aviv: Babel Press.

Rothwell, N. (2006, March 4) Scams in the desert. *The Australian*, pp. 19–22.

Rowse, T. (1998) *White Flour, White Power: From rations to citizenship in Central Australia*. New York: Cambridge University Press.

Rubuntja, W. (2002) *The Town Grew up Dancing: The art and life of Wenten Rubuntja*. With Jenny Green and with contributions from Tim Rowse. Alice Springs: Jukurrpa Books.

Rutheiser, C. (1996) *Imagineering Atlanta: The politics of place in the city of dreams*. London: Verso.

Rydell, R. W. and Schiavo, L. B. (Eds.) (2010) *Designing Tomorrow: America's World Fairs of the 1930s*. New Haven, CT: Yale University Press.

Sanders, W. (2004) *Indigenous People in the Alice Springs Town Camps: The 2001 census data*. Center for Aboriginal Economic Policy Research. Discussion Paper No. 260/2004. Canberra: Australian National University.

Sanni, L. (2007) Lagos population is 17.5 million. *The Daily Independent*. Retrieved February 16, 2007 from www.independentngonline.com/news/44/ARTICLE/20419/2007–02–06.html.

Sarlo, B. (2001) Forgetting Benjamin. *Cultural Critique*, 49 (Autumn), 84–92.

Sassen, S. (2006) *Territory, Authority, Rights: From medieval to global assemblages*. Princeton, NJ: Princeton University Press.

Schell, O. (2007) China, the Olympics and global leadership. *Culture and the Olympics*, 9, 18–20.

Schillmeier, Michael (2009) Assembling money and the senses: Revisiting George Simmel and the city. In I. Farias and T. Bender (Eds.) *Urban Assemblages*. London: Routledge, pp. 229–252.

Schwartz, J. (2003, October 5) Imagining central Texas, 20–40 years down the road. *Austin American Statesman*, p. A1.

Scott, A. J. (2000) *The Cultural Economy of Cities*. London: Sage.

Scott, A. J. (Ed.) (2001) *Global City Regions*. Oxford, UK: Oxford University Press.

Searle, G. (2002) Uncertain legacy: Sydney's Olympic stadiums. *European Planning Studies*, 10, 845–860.

Segbers, K. (Ed.) (2007) *The Making of Global City Regions*. Baltimore, MD: Johns Hopkins University Press.

Sheller, M. and Urry, J. (2006) The new mobilities paradigm. *Environment and Planning A*, 38, 207–226.

Short, J. R. (1989) *The Humane City: Cities as if people matter*. Oxford, UK: Blackwell.

Short, J. R. (1991) *Imagined Country: Society, culture and environment*. London: Routledge.

Short, J. R. (1999) Urban imagineers: Boosterism and the representation of cities. In A. Jones and D. Wilson (Eds.) *The Urban Growth Machine: Critical perspectives two decades later*. Albany NY: SUNY Press, pp. 37–54.

Short, J. R. (2001) *Global Dimensions: Space, place and the contemporary world*. London: Reaktion.

Short, J. R. (2004a) *Global Metropolitan*. London: Routledge.

Short, J. R. (2004b) Black holes and loose connections in a global urban network. *Professional Geographer*, 56, 295–302.

Short, J. R. (2004c) *Making Space: Revisioning the world, 1475–1600*. Syracuse, NY: Syracuse University Press.

Short, J. R. (2006) *Alabaster Cities: Urban US since 1950*. Syracuse, NY: Syracuse University Press.

Short, J. R. (2007) *Liquid City: Megalopolis and the contemporary northeast*. Washington, DC: Resources For The Future.

Short, J. R. and Kim, Y-H. (1998) Urban crises/urban representations: Selling the city in difficult times. In T. Hall and P. Hubbard (Eds.) *The Entrepreneurial City: Geographies of politics, regime and representation*. Chichester: John Wiley, pp. 55–75.

Short, J. R. and Peralta, L. M. P. (2009) Urban primacy: reopening the debate. *Geography Compass*, 3, 1245–1266.

Short, J. R., Hanlon, B., and Vicino T. J. (2007) The decline of inner suburbs: The new suburban Gothic in the United States. *Geography Compass*, 1, 641–656,

Short, J. R., Kim, Y., Kuus, M., and Wells, H. (1996) The dirty little secret of world cities research. *International Journal of Urban and Regional Research*, 20, 697–717.

Shorter, C. (1908) *The Brontes' Life and Letters*. New York: Scribner.

Shoval, N. (2002) A new phase in the competition for the Olympic gold: The London and New York bids for the 2012 Games. *Journal of Urban Affairs*, 24, 583–599.

Simmel, G. (1997) The metropolis and mental life [1903]. In D. Frisby and M. Featherstone (Eds.) *Simmel on Culture*. London: Sage, pp. 243–255.

Smith, N. (1996) *The New Urban Frontier: Gentrification and the revanchist city*. London and New York: Routledge.

Spiridon, M. (2005) In the wake of Walter Benjamin. *Interlitteraria*, 10, 150–159.

Stewart, M. (2006) *The Courtier and the Heretic: Leibniz, Spinoza and the fate of god in the modern world*. New York: W. W. Norton.

Taeuber, I. and Taeuber, C. (1964) The great concentration: SMSA's from Boston to Washington. *Population Index*, 30, 3–29.

Taylor, P. J. (2004) *Global City Network*. London: Routledge.

Thompson, W. (2000) "The Symbol of Paris": Writing the Eiffel Tower. *The French Review*, 73, 1130–1140.

Toulmin, S. (1990) *Cosmopolis: The hidden agenda of modernity*. New York: Free Press.

U.S. Department of Health and Human Services, Health Resources and Services Administration, Maternal and Child Health Bureau (2005) *Child Health USA 2005*. Rockville, MD: U.S. Department of Health and Human Services.

Varadarajan, S. (2007) Dictatorship by cartography. Retrieved July 2, 2010 from http:// svaradarajan.blogspot.com/2007/02/dictatorship-by-cartography-geometry.html.

Vicino, T. J., Hanlon, B. and Short, J. R. (2007) Megalopolis 50 years on: The transformation of a city region. *International Journal of Urban and Regional Research*, 31, 344–367.

Virilio, P. (1986) *Speed and Politics*. New York: Semiotext.

Virilio, P. (1995) *The Art of the Motor*. New York: Semiotext.

Virilio, P. (2005) *Negative Horizon: An essay in dromoscopy*. London: Continuum.

Waitt, G. (1999) Naturalizing the "primitive": A critique of marketing Australia's indigenous people as "hunter-gatherers." *Tourism Geographies*, 1, 142–163.

Waitt, G. (2008) Urban festivals: Geographies of hype, helplessness and hope. *Geography Compass*, 2, 513–537.

Wallerstein, I. (1974) *The Modern World System. Vol. 1*. New York: Academic Press.

Wallerstein, I. (1980) *The Modern World System. Vol. 2*. New York: Academic Press.

Wallerstein, I. (1989) *The Modern World System. Vol. 3*. San Diego, CA: Academic Press.

Wang, J. and Lau, S. Y. (2008) Forming foreign enclaves in Shanghai: State action in globalization. *Journal of Housing and the Built Environment*, 23, 103–108.

Wang, W. (2009) New trinity in today's Shanghai: Real estate market as an example. In *Humanistic Reflections for the Humane City*. Incheon: Incheon Conference for Urban Humanities, pp. 89–116.

Wasserstrom, J. N. (2009) *Global Shanghai, 1850–2010*. London: Routledge.

Watson, C. (2003) *Piercing the Ground*. Freemantle: Fremantle Arts Center Press.

Watts, M. (2005) Baudelaire over Berea, Simmel over Sandton? *Public Culture*, 17, 181–192.

Wei, Y. D. and Leung, C. K. (2005) Development zones, foreign investment and global city formation in Shanghai. *Growth and Change*, 36, 16–40.

Weller, R. H. (1967) An empirical examination of megalopolitan structure. *Demography*, 4, 734–743.

White, D., Raeside, R., and Barker, D. (2000) *Road Accidents and Children Living in Disadvantaged Areas*. Napier University: The Scottish Executive Central Research Unit.

Wilk, C. (Ed.) (2006) *Modernism 1914–1939: Designing a New World*. London: V & A Publications.

Williams, R. (1976) *Keywords*. London: Fontana.

Wilson, H. (1996) What is an Olympic city? Visions of Sydney 2000. *Media, Culture & Society*, 18, 603–618.

Wittels, D. (1949, November 12) They asked to be killed. In B. Hibbs (Ed.) *The Saturday Evening Post,* 65–66.

Wolff, J. (2000) The feminine in modern art: Benjamin, Simmel and the gender of modernity. *Theory, Culture & Society*, 17, 33–53.

Wolin, R. (1982) *Walter Benjamin: An aesthetic of redemption.* New York: Columbia University.

Wong, T-C., Yeow, M-C. and Zhu, X. (2005) Building a global city: Negotiating the massive influx of floating population in Shanghai. *Journal of Housing and the Built Environment*, 20, 21–50.

World Bank (2009) *World Development Report 2009: Reshaping economic geography.* Washington, DC.

World Markets Country Analysis (2002). Retrieved March 5, 2002 from www.world marketanalysis.com.

Wu, F. (2003) Globalization, place promotion and urban development in Shanghai. *Journal of Urban Affairs*, 25, 55–78.

Wu, F., Xu, J., and Gar-On Yeh, A. (2007) *Urban Development in Post-reform China: State, market and space.* London: Routledge.

Wu, W. (2008) Migrant settlement and spatial distribution in metropolitan Shanghai. *Professional Geographer*, 60, 101–120.

Wunderlich, F. M. (2008) Walking as rhythmicity: Sensing urban space. *Journal of Urban Design*, 13, 125–139.

Wyly, E. K. and Hammel, D. J. (2004) Gentrification, segregation and discrimination in the American urban system. *Environment and Planning A*, 36, 1, 215–1, 241.

Xu, X. (2006) Modernizing China in the Olympic spotlight: China's national identity and the 2008 Beijing Olympiad. *Sociological Review*, 54, 90–107.

Yee, W-S. (2007) *Shanghai Splendor: Economic sentiment and the making of modern China, 1843–1949.* Berkeley and Los Angeles, CA: University of California Press.

Yeoh, B. S. A. (2001) Postcolonial cities. *Progress in Human Geography*, 25, 456–468.

Yue, M. (2006) *Shanghai and the Edges of Empire.* Minneapolis, MN: University of Minnesota Press.

Index

Page numbers in italic type indicate relevant figures and tables.